AI SNAKE OIL

AI SNAKE OIL

**What Artificial Intelligence
Can Do, What It Can't,
and How to Tell the Difference**

ARVIND NARAYANAN
& SAYASH KAPOOR

With a new preface and a new epilogue
by the authors

PRINCETON UNIVERSITY PRESS

PRINCETON & OXFORD

Copyright © 2024 by Princeton University Press

Preface and epilogue to the paperback edition copyright © 2025 by
Princeton University Press

Princeton University Press is committed to the protection of copyright
and the intellectual property our authors entrust to us. Copyright
promotes the progress and integrity of knowledge created by humans.
By engaging with an authorized copy of this work, you are supporting
creators and the global exchange of ideas. As this work is protected by
copyright, any reproduction or distribution of it in any form for any
purpose requires permission; permission requests should be sent to
permissions@press.princeton.edu. Ingestion of any IP for any AI
purposes is strictly prohibited.

Published by Princeton University Press
41 William Street, Princeton, New Jersey 08540
99 Banbury Road, Oxford OX2 6JX

press.princeton.edu

GPSR Authorized Representative: Easy Access System Europe - Mustamäe
tee 50, 10621 Tallinn, Estonia, gpsr.requests@easproject.com

All Rights Reserved

First paperback printing, 2025
Paperback ISBN 9780691249148
ISBN (e-book) 9780691277929
Library of Congress Control Number: 2025933318

British Library Cataloging-in-Publication Data is available

Editorial: Hallie Stebbins, Chloe Coy
Jacket/Cover: Karl Spurzem
Production Editorial: Elizabeth Byrd
Production: Erin Suydam
Publicity: Maria Whelan (US); Kate Farquhar-Thomson (UK)
Copyeditor: Kelly Walters

This book has been composed in Arno Pro

Printed in the United States of America

To my wife, Veena
—*Arvind*

For Vineeta Kapoor and Ravi Kapoor,
my first mentors, writing instructors, editors,
and so much more
—*Sayash*

CONTENTS

Preface to the Paperback Edition   xi

1 Introduction   1

The Dawn of AI as a Consumer Product   3

AI Shakes Up Entertainment   7

Predictive AI: An Extraordinary Claim That Requires Extraordinary Evidence   9

Painting AI with a Single Brush Is Tempting but Flawed   12

A Series of Curious Circumstances Led to This Book   18

The AI Hype Vortex   21

What Is AI Snake Oil?   26

Who This Book Is For   34

2 How Predictive AI Goes Wrong   36

Predictive AI Makes Life-Altering Decisions   38

A Good Prediction Is Not a Good Decision   43

Opaque AI Incentivizes Gaming   46

Overautomation   48

Predictions about the Wrong People   51

Predictive AI Exacerbates Existing Inequalities	53
A World without Prediction	56
Concluding Thoughts	58

3 Why Can't AI Predict the Future?   60

A Brief History of Predicting the Future Using Computers	62
Getting Specific	67
The Fragile Families Challenge	70
Why Did the Fragile Families Challenge End in Disappointment?	73
Predictions in Criminal Justice	78
Failure Is Hard. What about Success?	81
The Meme Lottery	86
From Individuals to Aggregates	90
Recap: Reasons for Limits to Prediction	97

4 The Long Road to Generative AI   99

Generative AI Is Built on a Long Series of Innovations Dating Back Eighty Years	105
Failure and Revival	107
Training Machines to "See"	111
The Technical and Cultural Significance of ImageNet	114
Classifying and Generating Images	118
Generative AI Appropriates Creative Labor	122
AI for Image Classification Can Quickly Become AI for Surveillance	127

	From Images to Text	129
	From Models to Chatbots	133
	Automating Bullshit	139
	Deepfakes, Fraud, and Other Malicious Uses	142
	The Cost of Improvement	143
	Taking Stock	146
5	**Is Advanced AI an Existential Threat?**	150
	What Do the Experts Think?	151
	The Ladder of Generality	156
	What's Next on the Ladder?	162
	Accelerating Progress?	165
	Rogue AI?	168
	A Global Ban on Powerful AI?	172
	A Better Approach: Defending against Specific Threats	174
	Concluding Thoughts	177
6	**Why Can't AI Fix Social Media?**	179
	When Everything Is Taken Out of Context	183
	Cultural Incompetence	188
	AI Excels at Predicting . . . the Past	194
	When AI Goes Up against Human Ingenuity	198
	A Matter of Life and Death	201
	Now Add Regulation into the Mix	205
	The Hard Part Is Drawing the Line	209

Recap: Seven Shortcomings of AI for Content Moderation	216
A Problem of Their Own Making	218
The Future of Content Moderation	223

7 Why Do Myths about AI Persist? — 227

AI Hype Is Different from Previous Technology Hype	231
The AI Community Has a Culture and History of Hype	235
Companies Have Few Incentives for Transparency	239
The Reproducibility Crisis in AI Research	241
News Media Misleads the Public	247
Public Figures Spread AI Hype	251
Cognitive Biases Lead Us Astray	255

8 Where Do We Go from Here? — 258

AI Snake Oil Is Appealing to Broken Institutions	261
Embracing Randomness	265
Regulation: Cutting through the False Dichotomy	268
Limitations of Regulation	274
AI and the Future of Work	276
Growing Up with AI in Kai's World	281
Growing Up with AI in Maya's World	285

Epilogue to the Paperback Edition   291
Acknowledgments   301
References   303
Index   343

PREFACE TO THE PAPERBACK EDITION

Readers of the hardcover edition of *AI Snake Oil* often told us they picked up the book expecting an anti-technology screed. On discovering it to be quite balanced, some were pleasantly surprised, a few others unpleasantly so. Given the title—and despite the more nuanced subtitle—we certainly can't blame any of them. So we thought it best to let you know up front what kind of book you're about to read.

To be clear, we don't hesitate to call a spade a spade and a sham a sham in the book. But we try hard to avoid sweeping generalities about AI. After all, if we thought all AI was useless, we could have saved you the trouble of reading a whole book. And it is precisely because some AI applications have improved rapidly that AI hype has been so misleading. When we think of AI as a single technology (it isn't), it is easy to forget that its usefulness varies greatly depending on the application, task, and context.

Speaking of rapid advances, we're often asked how long a book on AI can possibly stay relevant. Fortunately, the goal of this book isn't to comment on breaking developments, but to give you foundational knowledge that will help you separate genuine advances from hype. The book has been out long enough that we can look back and evaluate how well we did. Happily, we've found that there is nothing in the text that needs to be corrected as a result of advancing technology, so we are cautiously optimistic that it will continue to stand the test of time.

The success of books also requires the *perception* of relevance. In this regard, we were fortunate in terms of timing. When ChatGPT was released, it was widely seen as a dramatic leap in AI. In reality, as we explain in the book, the underlying technology has an eighty-year history and had been advancing gradually; ChatGPT was only a minor improvement over its predecessor, GPT-3. If there was a step change, it was in public perception, not AI capability. If this book had been written shortly before ChatGPT was released, instead of after, it would have made the same fundamental points. But a book that didn't mention the product, let alone discuss it at length, would have been a complete flop, because people would have mistakenly assumed that it couldn't possibly be relevant.

We write this preface to the paperback edition with the same trepidation—that there will be a new product tomorrow that is portrayed as having revolutionized AI. If that happens, don't be fooled. It probably isn't a revolution.

The reason why the book's core arguments aren't threatened by advances in AI is that most of the limitations of AI that we point out aren't technical. Consider predictive AI, which is used in hiring, education, criminal justice, and many other domains to make life-altering decisions about people by predicting their future behavior. We are generally skeptical about predictive AI for two main reasons. One reason is that it's hard to predict the future, regardless of whether AI is used. The other is that it often represents a mindset of technocratic social control that ends up harming the people subjected to automated decisions. These arguments have much more to do with human nature and societies than with the technology in question.

In the book we analogize some kinds of predictive AI to elaborate random number generators. People often ask us if it's so bad to use a random number generator to make decisions. It isn't! In fact, we advocate for partial lotteries in many settings such as hiring and college admissions. In a partial lottery, applicants are selected randomly from among those who meet certain criteria. But right now, the idea enjoys very little support with the public. The appeal of predictive AI is based on the collective delusion that AI is good at prediction. If our book helps change this attitude even a little, and gets people to recognize that we often have to live with randomness, we'd consider it a success. Moving from predictive AI to partial lotteries would lead to decisions that are no worse, and would have the huge benefit of avoiding the shifting of power from decision makers and decision subjects toward unaccountable AI vendors.

On the other hand, we are cautiously optimistic about generative AI applications such as chatbots. Interestingly, one area of ongoing technical change is that the technology behind chatbots is being applied to predictive AI. This doesn't challenge the categorization of AI that guides our thinking and prescriptions. That's because the distinction that matters most to us is the application, not the underlying technology (the book isn't adequately clear about this). If a company asks a chatbot to read résumés and predict who will perform well at a job, and accepts or rejects job applicants on the basis of the responses, that's still predictive AI, and all of our skepticism applies. The point of the application is to make decisions, not generate text.

Turning to actual generative AI, we say in the book that it will be useful in some fashion to most people whose work is cognitive in nature. We are heavy users of generative AI ourselves, especially for computer programming, an area where capabilities are rapidly advancing. AI can generate snippets of code, and even entire applications, albeit simple ones, based on a text description of what the code should do. In some cases we've even found that it's faster to build an app with AI to do what you want than to look for an app that suits your needs in an app store. For many programmers, including us, it's hard to imagine going back to an era before the availability of AI assistance for writing code.

On the other hand, the difficulty of avoiding pitfalls has also been increasing over time as generative AI has become much more capable and products have proliferated. For example, there can be subtle bugs in AI-generated code. For now, it is *caveat emptor*. This book will give you a basic understanding of what generative AI's risks and limitations are, but it takes time and experimentation to develop a working understanding that is tailored to the specific ways you might want to use AI in your workflows. Given that there can be a significant learning curve, is generative AI still worth it from a productivity perspective? Well, sometimes. It's hard to offer general guidance.

That said, productivity isn't the only reason to use generative AI for work. The book didn't talk about how much fun it can be. We often find ourselves turning to generative AI to make work more fun—sometimes to automate mundane tasks, sometimes as a "thinking partner" that enhances our creativity. We encourage you to explore it for this reason alone.

Sadly, the impact of using generative AI on work satisfaction is not uniform across fields. In some jobs, people have reported that AI automates the interesting and creative parts, leaving the drudgery to humans. When such conflicts between efficiency and other values arise, whether or not to accept the tradeoff is up to us. Unfortunately, in a capitalist system, individuals might not be able to make this choice, so collective action may be necessary. The book gives some examples of organized groups of people successfully resisting uses of AI that would have negatively affected their lives or careers.

In fact, capitalism is a theme that pervades the book. Unsurprisingly, reactions to the book seem to be strongly influenced by readers' stances on capitalism. We think it's a necessary but flawed system, and that constant societal attention and action are required in order to minimize the harms and inequalities that it creates. Our hope with the book is to contribute to our collective capacity for such attention and action, and we are grateful that the book seems to resonate strongly with readers who share our underlying views.

Readers who have fewer reservations about capitalism were unconvinced by parts of the book, such as our call for more worker rights, and don't see the problem with the enormous power asymmetry between huge AI companies and others such as journalists, artists, and the workers in developing countries who are employed to train AI under often traumatic working conditions. Such disagreements reflect profound differences in worldviews and aren't primarily about AI. In general, if one's views on what makes a just society are entirely constrained by what is possible within existing markets, then there is never any justification for significant change.

On the other hand, some readers felt that we weren't critical enough of capitalism, big tech companies, and other powerful institutions. Those who view digital technology as fundamentally a force of oppression in capitalistic societies are likely to find our book disappointing and not radical enough.

There is plenty of greed in the stories we tell, but there aren't villains to conveniently place blame on, whether individuals, companies, or institutions. For example, while we're strongly critical of AI doom narratives from the industry that ironically only further feed into AI hype, we think most proponents of these narratives are sincere in their beliefs.

Even in the most seemingly straightforward cases of crooked AI companies, there is usually more than meets the eye. Many startups have been found to be selling AI products or services that turned out not to be automated—humans did the work behind the screens. It's unethical, to be sure, and possibly unlawful to lie about one's products. But one way in which the lie tends to come about is this: The company raises money to build an AI product, hires humans to train the AI, launches the product prematurely, and belatedly finds that it is harder than expected to get the AI to work well enough. Admitting that it has been using humans who are supposed to be training AI to actually provide the service would spook investors, so the company decides to keep it quiet. Our point here is not to excuse AI snake oil, but to make it more understandable.

Sometimes these underlying factors may not be apparent from the way AI failures are reported. Epic is a healthcare technology company that sold a flawed sepsis detection AI tool to hospitals. In our telling, based on the publicly available information, the blame lay squarely with the company. But we heard from a doctor who read our book that Epic tends to be very transparent with its customers—unusually so—enabling hospitals to do their own testing of its tools. Yet most hospitals don't, due to a lack of expertise or resources.

In short, most AI snake oil arises because of a complex interplay between various actors pursuing their own incentives. There are no easy fixes. On the spectrum of responses between incremental and radical—rooting out bad apples, reforming institutions, overthrowing institutions—this book advocates a middle ground.

AI SNAKE OIL

Chapter 1

INTRODUCTION

IMAGINE AN ALTERNATE universe in which people don't have words for different forms of transportation—only the collective noun "vehicle." They use that word to refer to cars, buses, bikes, spacecraft, and all other ways of getting from place A to place B. Conversations in this world are confusing. There are furious debates about whether or not vehicles are environmentally friendly, even though no one realizes that one side of the debate is talking about bikes and the other side is talking about trucks. There is a breakthrough in rocketry, but the media focuses on how vehicles have gotten faster—so people call their car dealer (oops, vehicle dealer) to ask when faster models will be available. Meanwhile, fraudsters have capitalized on the fact that consumers don't know what to believe when it comes to vehicle technology, so scams are rampant in the vehicle sector.

Now replace the word "vehicle" with "artificial intelligence," and we have a pretty good description of the world we live in.

Artificial intelligence, AI for short, is an umbrella term for a set of loosely related technologies. ChatGPT has little in common with, say, software that banks use to evaluate loan applicants. Both are referred to as AI, but in all the ways that matter—how they work, what they're used for and by whom, and how they fail—they couldn't be more different.

Chatbots, as well as image generators like Dall-E, Stable Diffusion, and Midjourney, fall under the banner of what's called generative AI. Generative AI can generate many types of content in seconds: chatbots generate often-realistic answers to human prompts, and image generators produce photorealistic images matching almost any description, say "a cow in a kitchen wearing a pink sweater." Other apps can generate speech or even music.

Generative AI technology has been rapidly advancing, its progress genuine and remarkable. But as a product, it is still immature, unreliable, and prone to misuse. At the same time, its popularization has been accompanied by hype, fear, and misinformation.

In contrast to generative AI is predictive AI, which makes predictions about the future in order to guide decision-making in the present. In policing, AI might predict "How many crimes will occur tomorrow in this area?" In inventory management, "How likely is this piece of machinery to fail in the next month?" In hiring, "How well will this candidate perform if hired for this job?"

Predictive AI is currently used by both companies and governments, but that doesn't mean it works. It's hard to predict the future, and AI doesn't change this fact. Sure, AI can be used to pore over data to identify broad statistical patterns—for instance, people who have jobs are more likely to pay back loans—and that can be useful. The problem is that predictive AI is often sold as far more than that, and it is used to make decisions about people's lives and careers. It is in this arena that most AI snake oil is concentrated.

AI snake oil is AI that does not and cannot work as advertised. Since AI refers to a vast array of technologies and applications, most people cannot yet fluently distinguish which types of AI are actually capable of functioning as promised and which types are simply snake oil. This is a major societal problem: we need to be able to separate the wheat from the chaff if we are to make full use of what AI has to offer while protecting ourselves from its possible harms, harms which in many cases are already occurring.

This book is a guide to identifying AI snake oil and AI hype. In it, we'll give you essential vocabulary to tease apart generative AI, predictive AI, and other types of AI. We'll share common-sense ways of assessing whether or not a purported advance is plausible. This will make you read news about AI much more skeptically and with an eye toward details that often get buried. A deeper understanding of AI will both satisfy your scientific curiosity and translate into practical ideas on how to use—and when not to use—AI in your life and career. And we will make the argument that predictive AI not only does not work today but will likely never work, because of the inherent difficulties in predicting human behavior. Finally, we hope that this book will get you thinking about your own responsibilities—and opportunities for change—with respect to the harmful implications of these tools.

The Dawn of AI as a Consumer Product

Generative AI burst into public view with the release of Chat-GPT in November 2022. Ironically, the developer OpenAI released it without any fanfare and internally thought of it as a "research preview."[1] But it went viral overnight as people shared amusing and amazing examples of its capabilities. For example, software developer Thomas Ptacek asked it to write a biblical verse in the style of the King James Bible, explaining how to remove a peanut butter sandwich from a VCR.[2] This is what ChatGPT spat out:

> And it came to pass that a man was troubled by a peanut butter sandwich, for it had been placed within his VCR, and he knew not how to remove it.
>
> And he cried out to the Lord, saying, "Oh, Lord, how can I remove this sandwich from my VCR, for it is stuck fast and will not budge?"
>
> And the Lord spoke unto him, saying, "Fear not, my child, for I shall guide thy hand and show thee the way. Take thy butter knife, and carefully insert it between the sandwich and the VCR, and gently pry them apart. And with patience and perseverance, the sandwich shall be removed, and thy VCR shall be saved."
>
> And the man did as the Lord commanded, and lo and behold, the sandwich was removed from the VCR, and the man was saved.
>
> And the Lord said, "Verily I say unto thee, seek not to put thy peanut butter sandwiches in thy VCR, for it is not a suitable place for such things. Rather, keep thy sandwiches in thy refrigerator or on thy plate, where they belong."
>
> And the man heeded the Lord's words, and from that day forth, he kept his sandwiches in their proper place, and was saved from trouble and woe. Amen.

Two months later, the app reportedly had over one hundred million users.[3] OpenAI was caught off guard by the explosion of interest and hadn't even procured enough computing power to handle the traffic that it generated.

Computer programmers soon started using it, because it turned out that ChatGPT was pretty good at generating snippets of programming code given only a description of what the code should do. In fact, programmers had been using a previous product called GitHub Copilot that used similar technology, but their use of AI accelerated after ChatGPT's release. It decreased the time it took to create apps. Even nonprogrammers could now build some simple apps.

Microsoft soon licensed the technology from OpenAI, and its Bing search engine rolled out a chatbot that could answer questions based on search results. Google had built its own chatbot in 2021 but had not yet released it or integrated it into its products.[4] Bing's move was seen as an existential threat to Google, and Google hurriedly announced its own search chatbot called Bard (later renamed Gemini).

That's when things started to go wrong. In the promotional video for Bard, the bot said that the James Webb Space Telescope took the first picture of a planet outside the solar system. An astrophysicist pointed out that this was wrong.[5] Apparently Google couldn't get even a cherry-picked example right. Its market value instantly took a hundred-*billion*-dollar dip. That's because investors were spooked by the prospect of a search engine that would get much worse at answering simple factual queries if Google were to integrate Bard into search, as it had promised.[6]

Google's embarrassment, while expensive, was only a ripple that portended the wave of problems that arose from chatbots' difficulties with factual information. Their weakness is a consequence of the way they are built. They learn statistical patterns from their training data—which comes largely from the web—and then generate remixed text based on those patterns. But they don't necessarily remember what's in their training data. We'll dive into this in chapter 4.

Misuse of the technology is rampant. News websites have been caught publishing error-filled AI-generated stories on important topics such as financial advice, and then refusing to stop using the technology even after the errors came to light.[7] Amazon is overrun with AI-generated books, including a few mushroom foraging guides, where errors can be fatal if a reader trusts the book.[8]

It's easy to look at all the flaws and misuses of chatbots and conclude that the world has gone mad for being so gaga about a technology that is so failure prone. But that conclusion would be too simplistic.

We think most knowledge industries can benefit from chatbots in some way. We use them ourselves for research assistance, for tasks ranging from mundane ones such as formatting citations correctly, to things we wouldn't otherwise be able to do such as understanding a jargon-filled paper in a research area we aren't familiar with.

The catch is that it takes effort and practice to use chatbots while avoiding their ever-present pitfalls. But *inappropriate* uses are much easier, because someone trying to make a quick buck, say by selling an AI-generated book, doesn't often care if the contents are garbage. That's what makes chatbots so conducive to misuse.

There are thornier questions about power. Suppose web search companies replace their traditional list of ten links with AI-generated ready answers. Even assuming that accuracy problems are fixed, the result is basically a machine for rewriting content found on other websites and passing it off as original, without having to send traffic or revenue to those websites. If search engines simply presented others' content as their own, they would run afoul of copyright law. But AI-generated answers seem to skirt this issue, although there are many lawsuits seeking to change this as of 2024.[9]

AI Shakes Up Entertainment

Another generative AI technology that has captivated people is text-to-image generation. In mid-2023, it was estimated that over a billion images had been created using Dall-E 2 by OpenAI, Firefly by Adobe, and Midjourney (by a company of the same name).[10] Another widely used image generator is Stable Diffusion by Stability AI, which is openly available, meaning that anyone can modify it to their liking. Stable Diffusion–based tools have been *downloaded* over two hundred million times. Since users run it on their own devices, there is no central tally of how many images have been generated using it, but it is likely to be several billion.

Image generators have enabled a deluge of entertainment.[11] Unlike traditional entertainment, these images are endlessly customizable to each user's interests. Some people delight in fantastic landscapes or cityscapes. Others enjoy images of historical figures in modern situations, or famous people doing things they wouldn't normally do, such as the Pope wearing a puffer jacket, dubbed "Balenciaga Pope." Fake trailers for various movies such as *Star Wars* in the highly recognizable style of Wes Anderson—symmetrical framing, pastel colors, whimsical sets—have proven popular.

It's not only hobbyists who are excited about image generators: entertainment apps are big business. Video game companies have created in-game characters that players can have a natural conversation with.[12] Many photo editing apps now have generative AI functionality. So, for example, you can ask such an app to add balloons to a picture of a birthday party.

AI was a major point of contention in the 2023 Hollywood strikes.[13] Actors worried that studios would be able to use existing footage of them to train AI tools capable of generating new videos based on a script—videos that looked like they featured the real actors whose images and videos the AI tools were trained on. In other words, studios would be able to capitalize on actors' likenesses and past labor in perpetuity, but without compensation.

While the strikes have ended, the underlying tensions between labor and capital are sure to resurface, especially as the technology advances.[14] Many companies are working on text-to-video generators, while others are working on automating script writing. The end result might not be as artistically complex or valuable, but that might not matter to studios looking to crank out a summer blockbuster.

In the long run, we think that a combination of technology and law can alleviate most of the problems we've described, as well as amplify the benefits. For example, there are many promising technical ideas to make chatbots less likely to fabricate information, while regulation can curb intentional misuses. But in the short term, adjusting to a world with generative AI is proving to be painful, as these tools are highly capable but unreliable. It's as if everyone in the world has been given the equivalent of a free buzzsaw.

It will take work to integrate AI appropriately into our lives. A good example is what's happening in schools and colleges, given that AI can generate essays and pass college exams. Let's be clear—AI is no threat to education, any more than the introduction of the calculator was.[15] With the right oversight, it can be a valuable learning tool. But to get there, teachers will have to overhaul their curricula, their teaching strategies, and their exams. At a well-funded institution such as Princeton, where we teach, this is an opportunity rather than a challenge. In fact, we encourage our students to use AI. But many others have been left scrambling as ChatGPT suddenly put a potential cheating tool in the hands of millions of students.

Will society be left perpetually reacting to new developments in generative AI? Or do we have the collective will to make structural changes that would allow us to spread out the highly uneven benefits and costs of new innovations, whatever they may be?

Predictive AI: An Extraordinary Claim That Requires Extraordinary Evidence

Generative AI creates many social costs and risks, especially in the short term. But we're cautiously optimistic about the potential of this type of AI to make people's lives better in the long run. Predictive AI is a different story.

In the last few years, applications of predictive AI to predict social outcomes have proliferated. Developers of these applications claim to be able to predict future outcomes about people, such as whether a defendant would go on to commit a future crime or whether a job applicant would do well at a job. In contrast to generative AI, predictive AI often does not work at all.[16]

People in the United States over the age of sixty-five are eligible to enroll in Medicare, a state-subsidized health insurance plan. To cut costs, Medicare providers have started using AI to predict how much time a patient will need to spend in a hospital.[17] These estimates are often incorrect. In one case, an eighty-five-year-old was evaluated as being ready to leave in seventeen days. But when the seventeen days passed, she was still in severe pain, and couldn't even push a walker without help. Still, based on the AI assessment, her insurance payments stopped. In cases like this, AI technology is often deployed with sensible intentions. For example, without predictive AI, nursing homes would be logically incentivized to house patients forever. But in many cases, the goals of the system as well as how it's deployed change over time. One can easily imagine how Medicare providers' use of AI may have started as a way to create a modicum of accountability for nursing homes, but then morphed into a way to squeeze pennies out of the system regardless of the human cost.

Similar stories are prevalent across domains. In hiring, many AI companies claim to be able to judge how warm, open, or kind someone is based on their body language, speech patterns, and other superficial features in a thirty-second video clip. Does this really work? And do these judgments actually predict job performance? Unfortunately, the companies making these claims have failed to release any verifiable evidence that their products are effective. And we have lots of evidence to the contrary, showing that it is extremely hard to predict individuals' life outcomes, as we'll see in chapter 3.

In 2013, Allstate, an insurance company, wanted to use predictive AI to determine insurance rates in the U.S. state of Maryland—so that the company could make more money without losing too many customers. It resulted in a "suckers list"—a list of people whose insurance rates increased dramatically compared to their earlier rates.[18] Seniors over the age of sixty-two were drastically overrepresented in this list, an example of automated discrimination. It is possible that seniors are less likely to shop around for better prices and that AI picked up on that pattern in the data. The new pricing would likely increase revenue for the insurance company, yet it is morally reprehensible. While Maryland refused Allstate's proposal to use this AI tool on the grounds that it was discriminatory, the company does use it in at least ten other U.S. states.*

If individuals object to AI in hiring, they can simply choose not to apply for jobs that engage AI to judge résumés. When predictive AI is used by governments, however, individuals have no choice but to comply. (That said, similar concerns also arise if many companies were to use the same AI to decide who to hire.) Many jurisdictions across the world use criminal risk prediction tools to decide whether defendants arrested for a crime should be released before their trial. Various biases of these systems have been documented: racial bias, gender bias, and ageism. But there's an even deeper problem: evidence suggests that these tools are only slightly more accurate than randomly guessing whether or not a defendant is "risky."

One reason for the low accuracy of these tools could be that data about certain important factors is not available. Consider three defendants who are identical in terms of the features that might be used by predictive AI to judge them: age, the number of past offenses, and the number of family members with criminal histories. These three defendants would be assigned the same risk score. However, in this example, one defendant is deeply remorseful, another has been wrongly arrested by the police, and the third is itching to finish the job. There is no good way for an AI tool to take these differences into account.

Another downside of predictive AI is that decision subjects have strong incentives to game the system. For example, AI was used to estimate how long the recipient of a kidney transplant

---

* Many of the examples in this book, like this one, are from the United States, simply because that is where we are based. However, the lessons we draw from these examples are intended to be broadly applicable.

would live after their transplant.[19] The logic was that people who had the longest to live after a transplant should be prioritized to receive kidneys. But the use of this prediction system would *disincentivize* patients with kidney issues to take care of their kidney function. That's because if their kidneys failed at a younger age, they would be more likely to get a transplant! Fortunately, the development of this system involved a deliberative process with participation by patients, doctors, and other stakeholders. So, the incentive misalignment was recognized and the use of predictive AI for kidney transplant matching was abandoned.

We'll see many more failures of predictive AI in chapters 2 and 3. Are things likely to improve over time? Unfortunately, we don't think so. Many of its flaws are inherent. For example, predictive AI is attractive because automation makes decision-making more efficient, but efficiency is exactly what results in a lack of accountability. We should be wary of predictive AI companies' claims unless they are accompanied by strong evidence.

Painting AI with a Single Brush Is Tempting but Flawed

Generative and predictive AI are two of the main types of AI. How many other types of AI are there? There is no way to answer that question, since there is no consensus about what is and isn't AI.

Here are three questions about how a computer system performs a task that may help us determine whether the label AI is appropriate. Each of these questions captures something about what we mean by AI, but none is a complete definition. First, does the task require creative effort or training for a human to perform? If yes, and the computer can perform it, it might be AI. This would explain why image generation, for example, qualifies as AI. To produce an image, humans need a certain amount of skill and practice, perhaps in the creative arts or in graphic design. But even *recognizing* what's in an image, say a cat or a teapot—a task that is trivial and automatic for humans—proved daunting to automate until the 2010s, yet object recognition has generally been labeled AI. Clearly, comparison to human intelligence is not the only relevant criterion.

Second, we can ask: Was the behavior of the system directly specified in code by the developer, or did it indirectly emerge, say by learning from examples or searching through a database? If the system's behavior emerged indirectly, it might qualify as AI. Learning from examples is called machine learning, which is a form of AI. This criterion helps explain why an insurance pricing formula, for example, might be considered AI if it was developed by having the computer analyze past claims data, but not if it was a direct result of an expert's knowledge, even if the actual rule was identical in both cases. Still, many manually programmed systems are nonetheless considered AI, such as some robot vacuum cleaners that avoid obstacles and walls.

A third criterion is whether the system makes decisions more or less autonomously and possesses some degree of flexibility and adaptability to the environment. If the answer is yes, the system might be considered AI. Autonomous driving is a good example—it is considered AI. But like the previous criteria, this criterion alone can't be considered a complete definition—we wouldn't call a traditional thermostat AI, one that contains no electronics. Its behavior rather arises from the simple principle of a metal expanding or contracting in response to changes in temperature and turning the flow of current on or off.

In the end, whether an application gets labeled AI is heavily influenced by historical usage, marketing, and other factors. We won't fret about the fact that there's no consistent definition.

That might seem surprising for a book about AI. But recall our overarching message: there's almost nothing one can say in one breath that applies to all types of AI. Most of our discussion in the book will be about specific types of AI, and as long as each type is clearly defined, we'll be on the same page.

There's a humorous AI definition that's worth mentioning, because it reveals an important point: "AI is whatever hasn't been done yet." In other words, once an application starts working reliably, it fades into the background and people take it for granted, so it's no longer thought of as AI. There are many examples: Robot vacuum cleaners like the Roomba. Autopilot in planes. Autocomplete on our phones. Handwriting recognition. Speech recognition. Spam filtering. Spell-check. Yes, there was a time when spell-check was considered a hard problem!

We think these tools are all wonderful. They quietly make our lives better. These are the kinds of AI we want more of. This book is about the types of AI that are problematic in some way, because you wouldn't want to read three hundred pages on the virtues of spell-check. But it's important to recognize that not all AI is problematic—far from it.

Some new AI technologies will hopefully one day come to be seen as mundane. Today, self-driving cars often make the news for accidents and fatalities.[20] But safe automated driving is ultimately a solvable problem, although one whose difficulty has repeatedly been underestimated. The bigger challenge for society might be the massive labor displacement that the technology will cause if it becomes widespread—millions of people drive trucks, taxis, or rideshare vehicles. Still, if the safety problem is solved and the necessary social and political adjustments are made, we may one day take self-driving cars for granted, like we do elevators today.

However, we think other types of AI, notably predictive AI, are unlikely to become normalized. Accurately predicting people's social behavior is not a solvable technology problem, and determining people's life chances on the basis of inherently faulty predictions will always be morally problematic.

For a more in-depth case study of why we must avoid sweeping generalizations about AI, consider facial recognition, an AI technology that has civil liberties advocates concerned. It has led to many false arrests in the United States—six, as we write this—all Black people. Should the use of facial recognition by police be discontinued because it is error prone and misidentifies Black people more often?

One fact that's easy to miss in this debate is that all the false arrests involved a cascading set of police failures, most of them human errors rather than technological. Robert Williams was arrested for shoplifting in part based on the testimony of a security contractor who wasn't even present at the time of the theft.[21] Randall Reid was arrested in Georgia for a shoplifting crime in Louisiana—a state he had never set foot in.[22] Porcha Woodruff was arrested based on a 2015 photo, despite the fact that a 2021 driver's license photo was available.[23] And so on.

Policing errors leading to the arrest of the wrong person happen every day, and will probably continue whether or not facial recognition is used.

Besides, police have made hundreds of thousands of facial recognition searches, so the error rate of the technology is minuscule.[24] In fact, the error rate dropped to 0.08 percent—a fifty-fold decrease between 2014 and 2020—according to studies by the National Institute of Standards and Technology.[25]

Facial recognition AI, if used correctly, tends to be accurate because there is little uncertainty or ambiguity in the task. Such AI is trained using vast databases of photos and labels that tell it whether or not any two photos represent the same person. So, given enough data and computational resources, it will learn the patterns that distinguish one face from another. Facial recognition is different from other facial analysis tasks such as gender identification or emotion recognition, which are far more error prone.[26,27] The crucial difference is that the information required to identify faces is present in the images themselves. Those other tasks involve guessing something about a person—their gender identity or emotional state—based on their face, which puts an inherent limit on their accuracy.

Civil rights advocates have often lumped together facial recognition with other error-prone technologies used in the criminal justice system, like those that predict the risk of crime—despite the fact that the two technologies have nothing in common and the fact that error rates differ by many orders of magnitude. (The majority of people who are labeled "high risk" by predictive AI do not in fact go on to commit another crime.)

The biggest danger of facial recognition arises from the fact that *it works really well*, so it can cause great harm in the hands of the wrong people. Kashmir Hill, in her book *Your Face Belongs to Us*, details many harmful ways in which it has been used.[28] For example, oppressive governments can and do use it to identify people in peaceful protests and retaliate against them.[29]

Facial recognition can also be abused by private companies. Madison Square Garden is a famous venue for sports events and concerts in New York City. In 2022, lawyer Nicolette Landi was denied entry to a Mariah Carey concert at the venue.[30] Her boyfriend had bought the nearly $400 tickets for her birthday. She was one of many lawyers turned away from various events at Madison Square Garden. The reason? The company that operates the venue had banned all lawyers who worked at firms that had sued it—even if they weren't responsible for the lawsuit, and even if they were longtime visitors with season tickets. The ban was enforced using facial recognition.

When critics oppose facial recognition on the basis that it doesn't work, they may simply try to shut it down or shame researchers who work on it. This approach misses out on the benefits that facial recognition has brought. For example, the Department of Homeland Security used it in a three-week operation to solve child exploitation cold cases based on photos or videos posted by abusers on social media.[31] It reportedly led to hundreds of identifications of children and abusers. Of course, there are more mundane benefits of facial recognition as well: unlocking our smartphones or easily organizing photos into albums based on who appears in them.

To be clear, even though facial recognition can be highly accurate when used correctly, it can easily fail in practice. For example, if used on grainy surveillance footage instead of clear photos, false matches are more likely. U.S. pharmacy chain Rite Aid used a flawed facial recognition system that led to employees wrongly accusing customers of theft. False matches happened thousands of times. The company tried its best to keep the system a secret. Fortunately, law enforcement agencies were paying attention. The Federal Trade Commission banned Rite Aid from using facial recognition for surveillance purposes for five years.[32]

To summarize, a nuanced approach to the double-edged nature of facial recognition would be to engage in vigorous democratic debate to identify which applications are appropriate, to resist inappropriate uses, and to develop guardrails to prevent abuse or misuse, whether by governments or private actors.

A Series of Curious Circumstances Led to This Book

In late 2019, a former researcher from an AI company reached out to Arvind out of the blue. The company is in the lucrative business of hiring automation—a business that is filled with snake oil, as we described above. The researcher explained that people at the company knew the tool wasn't very effective, in contrast to the company's marketing claims, but the company had suppressed internal efforts to investigate its accuracy.

Coincidentally, around the same time, Arvind was invited to give a public lecture at MIT. The meeting with the researcher fresh in his mind, he spoke about AI snake oil, showcasing the sketchiness of hiring automation. Encouraged by the audience's reaction, he shared his presentation slides online, thinking that a few scholars and activists might find them interesting. But the slides unexpectedly went viral. They were downloaded tens of thousands of times and his tweets about them were viewed two million times.

Once the shock wore off, it was clear to Arvind why the topic had touched a nerve. Most of us suspect that a lot of the AI around us is fake, but we don't have the vocabulary or the authority to question it.[33] After all, it's being peddled by supposed geniuses and trillion-dollar companies. But a computer science professor calling bullshit gave legitimacy to those doubts. It turned out to be the impetus that people needed to share their own skepticism.

Within two days, Arvind's inbox had forty to fifty invitations to turn the talk into an article or even a book. But he didn't think he understood the topic well enough to write a book. He didn't want to do it unless he had a book's worth of things to say, and he didn't want to simply trade on the popularity of the talk.

The second best way to understand a topic in a university is to take a course on it. The best way is to teach a course on it. So that's what Arvind did, teaming up with Princeton sociology professor Matthew Salganik. Matt had published many foundational pieces of research showing why it's hard to predict the future with AI. We'll see two of them in chapter 3. The course was called Limits to Prediction. Matt and Arvind invited the students in the course to conduct research. One of the students in the course was Sayash.

Sayash had just joined Princeton, having previously worked at Facebook. He ultimately decided to leave Facebook to obtain a PhD and pursue public-interest technology outside a tech company. He was accepted to a few computer science PhD programs. Accepted students are invited to visit the departments in person, to meet prospective collaborators and ask questions to judge whether they would be a good fit.

When visiting departments, PhD students are advised to ask questions of this sort: What is your style of advising? How much time do your students take off? What is your approach to work-life balance? These questions are important, and they can tell you how an advisor works, but not what they value and how they think. A far more revealing question is "What would you do if a tech company files a lawsuit against you?" The answer can tell you the advisor's stance on Big Tech, how they view the impact of their research, and what they would do in a crunch. It is also unusual enough that potential advisors wouldn't have prepared their answers in advance.

Sayash asked every potential advisor this question. It carried the element of surprise, yet the scenario it described was not completely unthinkable. When Arvind answered, "I would be glad if a company threatened to sue me for my research, because that means my work is having an impact," Sayash knew he had found the right program.

In the course on limits to prediction, students in the class were interested in predictive AI: in any and all attempts to predict the future using data, especially in social settings, ranging from civilizations to social media. Some interesting questions we looked at were: Can we predict geopolitical events such as election outcomes, recessions, or social movements? Can we predict which videos will go viral?

What we found was a graveyard of ambitious attempts to predict the future. The same fundamental roadblocks seemed to come up over and over, but since researchers in different disciplines rarely talk to each other, many scientific fields had independently rediscovered these limits. We were alarmed by the contrast between the weight of the evidence and the widespread perception that machine learning is a good tool for predicting the future.

The course included many case studies, including Google Flu Trends. This was a project that Google launched in 2008 to predict flu outbreaks by analyzing the search queries that its millions of users make every day. An increase in searches for flu-related terms could be indicative of an imminent outbreak. Google heavily promoted it as an example of AI and mass data collection used for social good. But within a few years, the accuracy of the predictions dropped precipitously. One reason was that it is hard to distinguish between media-driven panic searches and actual increases in flu activity. Another was that Google's own changes to its app changed people's search patterns in ways that weren't accounted for by the AI. Google Flu Trends ultimately ended up as a cautionary tale.[34] The lesson is that even in cases where it is possible to make somewhat accurate forecasts, it is very easy to get the details wrong.

Sayash found that the course confirmed his previous experiences at Facebook, where he saw how easy it was to make errors when building AI and to be overoptimistic about its efficacy. Errors could arise due to many subtle reasons and often weren't caught in testing, but only when AI was actually deployed to real users.[35] Sayash decided to choose the limits of AI as his research topic.

After four years of research, separately and together, we're ready to share what we've learned. But this book isn't just about sharing knowledge. AI is being used to make impactful decisions about us every day, so broken AI can and does wreck lives and careers. Of course, not all AI is snake oil—far from it—so the ability to distinguish genuine progress from hype is critical for all of us. Perhaps our book can help.

The AI Hype Vortex

Since we started working together, we've come to better appreciate why there is so much misinformation, misunderstanding, and mythology about AI. In short, we realized that the problem is so persistent because researchers, companies, and the media all contribute to it.

Let's start with an example from the research world. A 2023 paper claimed that machine learning could predict hit songs with 97 percent accuracy.[36] Music producers are always looking out for the next hit, so this finding would have been music to their ears. News outlets, including *Scientific American* and Axios, published pieces about how this "frightening accuracy" could revolutionize the music industry.[37,38] Earlier studies had found that it is hard to predict if a song will be successful in advance, so this paper seemed to describe a dramatic achievement.

Unfortunately for music producers, we found that the study's results were bogus.

The method presented in the paper exhibits one of the most common pitfalls in machine learning: data leakage. This means roughly that the tool is evaluated on the same, or similar, data that it is trained on, which leads to exaggerated estimates of accuracy. This is like teaching to the test—or worse, giving away the answers before an exam. We redid the analysis after fixing the error and found that machine learning performed no better than random guessing.

This is not an isolated example. Textbook errors in machine learning papers are shockingly common, especially when machine learning is used as an off-the-shelf tool by researchers not trained in computer science. For example, medical researchers may use it to predict diseases, social scientists to predict people's life outcomes, and political scientists to predict civil wars.

Systematic reviews of published research in many areas have found that the *majority* of machine-learning-based research that was re-examined turned out to be flawed.[39] The reason is not always nefarious; machine learning is inherently tricky, and it is extremely easy for researchers to fool themselves. Overall, research teams in more than a dozen fields have compiled evidence of widespread flaws in their own arenas, unaware that they were all part of a far-reaching credibility crisis in machine learning.

The more buzzy the research topic, the worse the quality seems to be. There are *thousands* of studies claiming to detect COVID-19 from chest x-rays and other imaging data. One systematic review looked at over four hundred papers, and concluded that *none* of them were of any clinical use because of flawed methods.[40] In over a dozen cases, the researchers used a training dataset where all the images of people with COVID-19 were from adults, and all the images of people without COVID-19 were from children. As a result, the AI they developed had merely learned to distinguish between adults and children, but the researchers mistakenly concluded that they had developed a COVID-19 detector.

We ourselves discovered flaws in many studies, mainly in the field of trying to predict civil wars (in short: it doesn't work). When we tried to publish a paper about an entire body of research being flawed, no journal was interested. It is notoriously hard to correct flaws in the scientific record. We eventually published our paper, but only after reframing it to be more palatable, as a guide to future researchers to avoid these pitfalls.

These days, when we find flawed machine learning papers, we don't even try to correct the record. The system doesn't work. In fact, in many fields, studies that fail attempts at replication by other research groups are cited *more* than those that replicate successfully.[41] The party line among scientists is that science "self-corrects," meaning that the normal process of science is sufficient to root out flawed research, but everything we've seen about the process suggests otherwise.

To be clear, incorrect machine learning claims in research papers usually don't result in broken AI products on the market. If a music producer tried to predict hits using a flawed method, they would quickly find out that it doesn't work. (Commercial AI snake oil usually results from companies knowingly selling AI that doesn't work, rather than they themselves being fooled.) Still, the ocean of scientific misinformation damages the public understanding of AI, because the media tends to trumpet every purported breakthrough.

There are rays of hope, though. In summer 2022, we organized a day-long online workshop to discuss the spate of flawed machine-learning-based science. To our surprise, hundreds of scientists showed up. Based on the workshop, we led a team of about twenty researchers across many disciplines to devise an intervention: a simple checklist that helps scientists better document how they use machine learning, which can help minimize errors and make them easier to spot when they do creep in.[42] It's still early days, and it remains to be seen if our intervention will be adopted. At any rate, scientific practice changes glacially, and it is likely that things will continue to get worse for a while before they get better.

Let's turn to companies. While overhyped research misleads the public, overhyped products lead to direct harm. To study this, we teamed up with colleagues Angelina Wang and Solon Barocas and investigated uses of predictive AI in industry and government.[43] We documented about fifty applications spanning criminal justice, healthcare, welfare allocation, finance, education, worker management, and marketing. Most of these deployments are recent. During the 2010s, predictive AI extended its tentacles into many spheres of life, judging us and determining our opportunities in life based on data covertly collected about us.

We realized that while vendors of these tools aggressively chase clients, they are rarely transparent about how well their products work, or if they work at all. Notably, as far as we know, no hiring automation company has ever published a peer-reviewed paper validating its predictive AI, or even allowed an external researcher to evaluate it. Two of the leading companies made a show of external audits: Pymetrics contracted with a leading research group from Northeastern University, and HireVue contracted a noted independent auditor. But in both cases, the researchers were allowed to analyze only whether the AI was biased with respect to race or gender, and not whether it worked. The companies cleverly used a concern about discrimination to their advantage. If what you have is an elaborate random number generator that works equally poorly for everyone, it's easy to make it unbiased!

Here, too, there are slivers of good news. Regulators are wising up to the fact that many predictive AI products don't work. In 2023, the U.S. Federal Trade Commission (FTC) warned companies that "we're not yet living in the realm of science fiction, where computers can generally make trustworthy predictions of human behavior. Your performance claims would be deceptive if they lack scientific support or if they apply only to certain types of users or under certain conditions."[44] The key word here is "deceptive"; the FTC is authorized by Congress to police deceptive practices by companies. We hope companies will heed that warning.

If researchers and companies kindle the sparks of hype, the media fans the flames. Every day we are bombarded with stories about purported AI breakthroughs. Many articles are just reworded press releases laundered as news.

Of course, with the media so reliant on clicks and newsrooms so cash strapped, this is no surprise. Still, there are systemic problems in the industry besides crumbling revenue. Many AI reporters practice what's called access journalism. They rely on maintaining good relationships with AI companies so that they can get access to interview subjects and advance product releases. That means not asking too many skeptical questions.

Claims of AI being conscious have proven particularly irresistible to the media. When a Google engineer claimed in June 2022 that the company's internal chatbot had become sentient (and faced "bigotry"), just about every publication ran with that headline.[45] The same thing happened when Bing's chatbot claimed to be sentient in early 2023. That's despite the fact that most AI researchers don't think there is any scientific basis for these claims.

There are many AI journalists who rise above the fray and do excellent investigative work. But so far they are a handful, constantly swimming against the tide. We've had the opportunity to discuss the hype problem with journalists and speak at a few journalism conferences. We learned about many ongoing efforts to improve the quality of tech journalism.

For example, the Pulitzer Center funds a network of journalists to work on "in-depth AI accountability stories that examine governments' and corporations' uses of predictive and surveillance technologies to guide decisions in policing, medicine, social welfare, the criminal justice system, hiring, and more."[46] Many notable investigations have resulted from this program, including one by Ari Sen and Derêka K. Bennett for the *Dallas Morning News*. Sen and Bennett looked into Social Sentinel, an AI product used by schools across the United States to scan students' social media posts, purportedly to identify safety threats, but often misused to surveil student protests.[47]

The Pulitzer Center fellowships support only ten journalists per year. In the long run, whether or not journalism can serve as a check against Big Tech's power will depend on whether funding models like these—that don't rely on clicks—can be scaled up.

AI experts have a responsibility to speak up against hype, whether it comes from researchers, companies, or the media. We are trying to do our part. In our newsletter, at AISnakeOil.com, we comment on new developments in AI and help readers separate the milk from the froth.[48]

What Is AI Snake Oil?

In the late nineteenth and early twentieth centuries, snake oil peddlers were rampant in America, selling miracle cures and health tonics under false pretenses. Figure 1.1 shows a typical

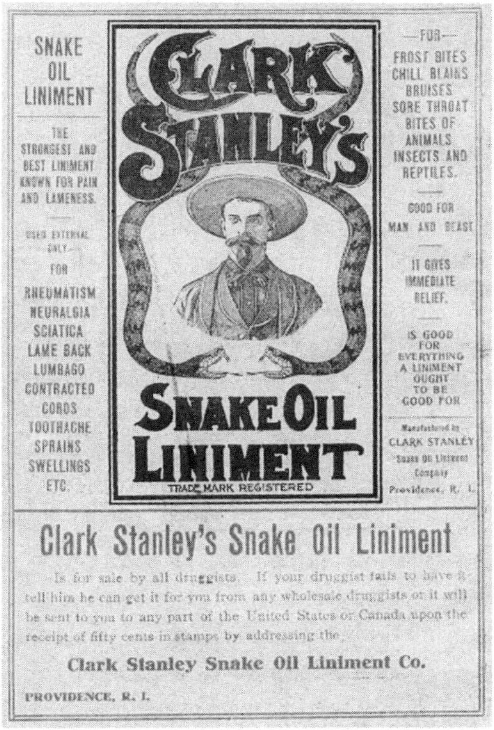

FIGURE 1.1. A 1905 advertisement for snake oil.
(*Sources:* https://www.nlm.nih.gov/exhibition/ephemera/medshow.html, attributed to Clark Stanley's Snake Oil Liniment, *True Life in the Far West*, 200 page pamphlet, illus., Worcester, Massachusetts, c. 1905, 23 × 14.8 cm. https://commons.wikimedia.org/w/index.php?curid=47338529.)

advertisement. Snake oil sellers exploited people's unscientific belief that oil from snakes had various health benefits, and their inability to tell effective treatments from useless ones. Besides, most of the concoctions being sold as snake oil didn't in fact contain any. In some cases, these medicines were ineffective but harmless. In others, they led to the loss of life or health. Until the Food and Drug Administration (FDA) was established in 1906, there was no good way to keep snake oil salesmen accountable to their promises regarding the contents, the efficacy, or the safety of their products.

AI snake oil is AI that does not and cannot work, like the hiring video analysis software that originally motivated the research that led to this book. The goal of this book is to identify AI snake oil—and to distinguish it from AI that can work well if used in the right ways. While some cases of snake oil are clear cut, the boundaries are a bit fuzzy. In many cases, AI works to some extent but is accompanied by exaggerated claims by the companies selling it. That hype leads to overreliance, such as using AI as a replacement for human expertise instead of as a way to augment it.

Just as important: even when AI works well, it can be harmful, as we saw in the example of facial recognition technology being abused for mass surveillance. To identify what the harm is and how to remedy it, it is vital to understand whether the problem has arisen due to AI failing to work, or being overhyped, or in fact working exactly as intended. Harm and truthfulness are the two axes in figure 1.2. In this book, we're interested in everything except the bottom left part of the figure, which is AI that both works and is benign.

With this picture in mind, here's a roadmap of the rest of the book.

Chapter 2 is about automated decision-making, which is one area where AI, specifically predictive AI, is increasingly used: predicting who will commit a crime, who will drop out of school, and so forth. We'll look at many examples of systems that have failed and caused great harm. In our research, we've identified a recurring set of reasons these failures keep happening—reasons that are intrinsic to the use of predictive

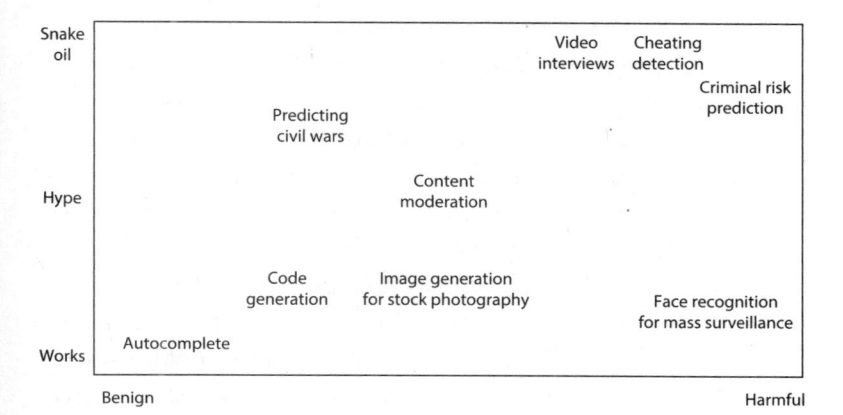

FIGURE 1.2. The landscape of AI snake oil, hype, and harms, showing a few illustrative applications.

logic in these high-impact systems. We'll end the chapter by asking if it is possible to reimagine decision-making without predictive AI, and we'll discuss what sorts of organizational and cultural adaptations we'll need in order to embrace the unpredictability inherent to consequential decisions.

In chapter 3 we'll take a step back to understand why predicting the future is so hard. Our answer is that its challenges are ultimately not about AI, but rather the nature of social processes; it is inherently hard to predict human behavior, and we'll see many reasons for this. We'll review evidence from many efforts to predict the future, from crime to children's life outcomes. We'll draw from academic studies as well as the rare cases where commercial products have been subjected to independent scrutiny. We'll look at prediction of both positive outcomes, such as succeeding at a job or publishing a bestseller, and negative outcomes, such as failing to pay back a loan; all of these turn out to be hard to predict. We'll also look at less consequential but more easily analyzed prediction tasks such as identifying which social media posts will go viral. And finally, in addition to outcomes about individuals, we'll look at macro-level predictions such as the evolution of pandemics. Across all of these domains, strikingly common patterns emerge, which lead us to conclude that the limitations of predictive AI won't go away in the foreseeable future.

It's simple to state the primary limitation of predictive AI: it's hard to predict the future. But with generative AI, to which we turn next, things are more complicated. The technology is remarkably capable, yet it struggles with many things a toddler can do.[49] It is also improving quickly. So, to understand what the limitations are and have some sense of where things might be going, it's important to understand the technology. In chapter 4, we hope to demystify how generative AI works.

We'll also discuss the many harms that arise from generative AI. In some cases, harms arise because the product is flawed. For instance, software that claims to detect AI-generated essays doesn't work, which can lead to false accusations of AI-based cheating. In other cases, harms arise because the product works well. Image generators are putting stock photographers out of jobs even as AI companies use their work without compensation to build the technology. Of course, there are many applications of generative AI that both work well and are broadly beneficial, such as automating some parts of computer programming (although, even here, there are minor risks that programmers should watch out for, such as the possibility of bugs in AI-generated code that might give hackers an advantage). Given the focus of the book, we won't spend much time on these beneficial applications. But we should emphasize that we are excited about them and about the potential of generative AI in general.

In chapter 5 we turn to questions of existential risk that have gripped the public discourse on AI. The fear is that once future AI systems are sufficiently advanced, they will be too hard to control. Our key perspective is that these fears rest on a binary notion of AI that crosses some critical threshold of autonomy or superhuman intelligence. But this idea is contradicted by the history of AI. The technology has gradually been increasing in flexibility and capability, which we explain by introducing the concept of a "ladder of generality." Current technology is already on the seventh rung of this ladder, each step being more general and powerful than the ones below it.

We have every reason to think that this pattern of step-by-step progress will continue. This means we don't have to speculate about the future but can instead learn from history. And what this grounded analysis shows is that claims of out-of-control AI rest on a series of flawed premises. Of course, we must take risks concerning powerful AI seriously. But we'll show that we already have the means to address them calmly and collectively.

In chapter 6, we turn to social media, where so-called recommendation algorithms are used for creating the personalized feeds we scroll through. AI is also used for determining which content violates policies and must be taken down; this process is called content moderation. The chapter is primarily about content moderation AI, with a brief discussion of recommendation algorithms. The central question we examine is whether AI has the potential to remove harmful content such as hate speech from social media without curbing free expression, as tech companies have often promised.

In this debate, much attention has been paid to the inevitable errors of enforcement, such as a piece of content being mistakenly flagged as unacceptable and taken down. But even if these errors are fixed, the more fundamental issue is that platforms have this power to regulate speech in the first place, with little accountability. We lack a democratic process to decide the rules by which online speech should be governed and to find a balance between values such as free speech and safety. Given this reality, AI will remain impotent at easing our frustrations with social media.

We have placed content moderation AI in the middle of figure 1.2. It works well enough that social media companies have come to rely on it, but it is often misleadingly portrayed as a way out of the moral and political quandaries that beset social media governance, which it is not. As for its harmfulness, while content moderation has often failed badly and even enabled large-scale real-world violence, we argue that these failures are not fundamentally about the technology but rather intrinsic consequences of handing over the digital public square to unaccountable private entities.

Predictive AI, generative AI, and content moderation AI are the three main types of AI that we'll discuss in this book. That's not an exhaustive list. As discussed earlier, there are many applications such as autocomplete or spell-check that work well and fly under the radar. There are also applications such as robotics and self-driving cars that are worth discussing but didn't make the cut, in part because they don't yet have societal consequences on the same scale as the applications we discuss. Still, the conceptual understanding we provide in this book will help you evaluate other AI applications as well—which ones are likely to work, and which ones might be snake oil.

In chapter 7 we consider the question of why myths about AI are so pervasive. Companies not only hype up their tech but also use their enormous wealth and power in ways that make academia and the press less effective as counterweights to their self-serving claims. In fact, academic researchers are as often a source of hype in AI as they are a voice of reason. In many fields, researchers have falsely come to a consensus that AI is highly accurate in their respective domains, all based on flawed and irreproducible research. We'll look at civil war prediction as one example. While faulty AI research usually doesn't lead to the deployment of faulty products—which is why we've put it in the top left quadrant of figure 1.2—it is nonetheless harmful and wasteful because it misleads the public. Turning to the media, we discuss ways in which journalists knowingly or unknowingly contribute to AI hype, and give you a guide for reading the news skeptically.

In the final chapter, we look at directions for change. We identify three broad paths. The first is to set ground rules for companies to govern how they build and advertise their products. We think there is an important role for regulation here, while we acknowledge that regulation shouldn't go overboard. The second path is the way in which we integrate AI into society. For example, what is the role of AI in education and in children's lives in general? And in the workplace, will we use AI to replace jobs or augment them? We see many of these questions as social and political choices rather than inevitable consequences of the technology itself.

Our third suggested path is to focus on the demand for AI snake oil rather than its supply. We show that over and over again, AI snake oil is appealing because those buying it are in broken institutions and are desperate for a quick fix. For example, schoolteachers, already overworked, have reacted badly to the disruption caused by students using AI to help with their homework. Unable to carry out the overhaul of their teaching and assessment strategies that AI necessitates, they have turned to cheating detection software. But these products don't work and have led to a spate of false accusations of academic dishonesty, often with disastrous consequences for students.

We can't fix these problems by fixing AI. If anything, AI snake oil does us a favor by shining a spotlight on these underlying problems. More broadly, we show how concerns about AI, especially in the labor market, are often really about capitalism. We must urgently figure out how to strengthen existing safety nets and develop new ones so that we can better absorb the shocks caused by rapid technological progress and reap its benefits.

Who This Book Is For

We hope this book will be interesting to three kinds of readers. Maybe you simply want to make sense of what's going on. You've seen the headlines, and you're wondering if AI can really predict earthquakes or pass the bar exam. And if so, how? Which jobs will still be around in twenty years? What will our children's lives be like?

What we offer to sate your curiosity are not philosophical musings about what it means to be human in the age of AI. Reasonable people can have different opinions on whether we're living through something profound or just the next step in the march of automation. Rather, what we hope to impart is a nuts-and-bolts understanding of what's going on behind the screen.

Or you may be interested in AI because you need to make decisions about AI at your workplace. We hope that this book will help you understand which types of AI work, which ones don't, and what the gotchas are. Throughout the history of AI, computer scientists have attempted to classify which problems are "easy" for AI and which ones are "hard." None of these sweeping generalizations has stood the test of time as the technology evolved. Our approach, instead, is to treat each kind of AI individually.

Finally, you may be interested in AI because you want to take action against the harms being perpetrated in the name of AI. Public-interest advocates have built up effective movements to resist harmful predictive AI. But with generative AI, the battle lines are still forming. If the technology leads to an economic transformation, as generative AI companies hope, it probably won't be good news for labor, whether or not it eliminates jobs. That's because this type of AI relies on the invisible, drudging, low-wage work of millions of people to create training data, as well as the use of data found on the web without credit or compensation to the writers, artists, and photographers who created them.

In the wake of the Industrial Revolution, millions of new jobs were created in factories and mines, with horrific working conditions. It took many decades to secure labor rights and improve workers' wages and safety. Similarly, there is a movement today to secure labor rights, human creativity, and dignity in the face of encroaching automation.[50] It is far from clear that this movement will succeed. It's up to all of us.

Finally, a quick note to instructors and students using this book in a course: we have exercises and other pedagogical materials on our website, AISnakeOil.com.[47]

Chapter 2

HOW PREDICTIVE AI GOES WRONG

IN 2015, administrators at Mount St. Mary's University, a private university in Maryland, USA, wanted to increase their university's retention rate—the proportion of admitted students who go on to graduate. The school conducted a survey to find which students were struggling. This might sound like a worthy goal; once the administration knows who is struggling, they can offer additional support to those students to help them adjust to college. But instead, the president suggested dismissing students who weren't doing well. He reasoned that if these students dropped out in the first few weeks of the semester, as opposed to later, they wouldn't count as "enrolled" and therefore wouldn't decrease the university's retention rate.

In a faculty meeting, the president said: "My short-term goal is to have 20–25 people leave by the 25th [of September]. This one thing will boost our retention 4–5%."[1] Professors raised objections by pointing out that it is hard to judge who will be successful in the first few weeks after entering college. "This is hard for you because you think of the students as cuddly bunnies, but you can't," the president responded. "You just have to drown the bunnies . . . put a Glock to their heads."

This is a startling example, but the fact is that many schools would like to predict which students are at risk of dropping out—some of them for reasons more aligned with students' well-being. An AI-based product called EAB Navigate could automate this process. In its marketing pitch to schools, EAB claimed: "The model will provide your school and its advisors with invaluable and otherwise unobtainable insight into your students' likelihood of academic success." Even if some schools might use this insight to pressure students to leave, others could conceivably use it to design interventions that might help students stay in school. But interventions that seem helpful could also be questionable. For example, the tool helps by recommending alternative majors in which a student would be more likely to succeed. This might have the effect of driving out poorer and Black students—whom the tool is more likely to flag—from more lucrative but more challenging STEM majors.[2] And throughout this process, students may have no idea that they are being evaluated using AI.

EAB Navigate is an example of an automated decision-making system that uses predictive AI. In this realm, there is a vast amount of AI snake oil.

Companies selling these tools make strong claims about their utility.[3] They are deployed widely by governments and private companies. Still, unlike generative AI applications like ChatGPT, predictive AI has largely flown under the radar when it comes to public interrogation. In many cases, including EAB Navigate, the decision subjects don't even know they are being automatically evaluated.

In this chapter, we will see how predictive AI goes wrong. While a full account of these failures would take up more than the entire length of this book,[4,5]* we will highlight common failures that are nevertheless hard to fix, underscoring the difficulty of making predictive AI work.

Before we dive in, let's look more closely at how these automated decisions are made.

Predictive AI Makes Life-Altering Decisions

Algorithms like EAB Navigate are everywhere, and they are used to automate consequential decisions about you—with or without your knowledge.[6] When you show up at a hospital, an algorithm could determine if you need to be admitted overnight or released the same day. When you apply for public benefits such as child welfare, an algorithm could determine if your application is valid or if you are trying to commit fraud. When you apply for a job, an algorithm could determine if your application will be considered by a recruiter or discarded. When you visit a beach, an algorithm could determine if the water is safe to swim in.

An algorithm is a set of steps or rules used to make a decision. Sometimes these rules are developed by people or institutions. For example, during the COVID-19 pandemic, the U.S. government sent stimulus checks to help citizens deal with the economic hardships they faced. Adults were paid USD 1,200; children were paid USD 500. Policymakers decided these rules. But once the rules were established, people's eligibility was automatically determined based on their past tax records.

---

* Two popular incident databases record failures of AI, including predictive AI, in the real world. The AI Incident Database has over 600 reports, and the AI, Algorithmic, and Automation Incidents and Controversies Repository has over 1,400 reports, as of 2024.

1. If the applicant is a U.S. citizen and eighteen or older, send a check for USD 1,200.
2. If they are a U.S. citizen and under eighteen: send a check for USD 500.
3. If they're not a U.S. citizen or earn more than USD 75,000 per year*: don't send a check.

This type of algorithm, in which the rules are developed manually but applied automatically, is commonly employed in the public sector, such as in allocating funds for public housing or welfare.

But increasingly, algorithms are used to *develop* the rules automatically from patterns in past data. For instance, while watching Netflix, if you rated *Forrest Gump* highly but not *The Shining*, a recommendation algorithm might predict that you'll enjoy drama films over horror films. Here, a Netflix employee is not manually creating a rule stating that users who liked *Forrest Gump* should be recommended other drama films. Instead, your ratings and the time you spend watching different types of films determine what films you're recommended next. Unlike the first kind of algorithm, the rules are both developed and applied automatically.

This kind of algorithm is called a model, which is a term you might encounter frequently. A model is usually a set of numbers that mathematically specify how the system should behave. These numbers may not be intelligible to a human, even to the developer of the system, unless the model is explicitly designed with that goal in mind. Models are created from data, or "trained," using a set of statistical techniques called machine learning.

---

* The actual algorithm was slightly more complex: the amount paid if a person earned more than USD 75,000 tapered off. And there were different rules for people with children.

These models can be much more consequential than deciding your next Friday night movie. They are used for allocating scarce resources, such as jobs or loans. They provide certain opportunities to people and foreclose other opportunities. This is what we call predictive AI: models used for decision-making based on predictions about the future, such as who will do well at a job or who will pay back a loan.

Take the example of criminal justice, where predictive AI is used for many kinds of decisions: Should an inmate get parole? What should happen to people who are arrested? Before trial, a judge needs to decide whether a defendant should be detained in jail, released on bail (and if so, what the bail amount should be), or released without any bail, but perhaps with other restrictions such as ankle monitors.

A lot rides on the answer.[7] Spending time in jail can ruin someone's life. They can lose their income in the short term— and even if they are released, especially if they have restrictions such as ankle monitors, it will still be much harder to find a job. They suffer from increased physical and mental illnesses, both due to stigmatization and because of poor conditions in jail. And a large fraction of the people in jail are simply there because they can't afford high bail amounts.

As a result, the criminal justice system disproportionately burdens the poor and leads to cycles of poverty and racial inequality. Almost half a million people are in U.S. jails at any given time without having been convicted of a crime.[8] Despite violent crime in the country going down by almost 50 percent, the number of people jailed in the last four decades has nearly doubled.[9,10]

Many U.S. states mandate the use of risk assessment tools to decide whether a person should be released or detained before trial, and predictive AI is often used. The tools produce two main risk scores. One is the risk that the defendant will commit a crime, especially a violent crime, if released. The other is the risk that they will fail to appear in court on the appointed date. In each case, defendants are labeled low, medium, or high risk. The tools try to calculate these scores based on certain features of the defendants.

Let's zoom in on one predictive AI product: COMPAS, which stands for Correctional Offender Management Profiling for Alternative Sanctions. COMPAS relies on defendants' answers to 137 questions to make decisions about them.[11] It includes questions about their past history of crime or failure to appear for court cases. It also includes questions about things individuals have little or no control over, such as how often their family members have been arrested, or whether their friends or acquaintances have been arrested. And some questions seem to punish people for their personality or for poverty, such as "How often do you feel bored?" and "How often do you have barely enough money to get by?"

The developers of COMPAS aim to predict if a defendant will fail to appear in court or be arrested for a crime within two years.[12] It is trained using data on defendants' behavior in the past. COMPAS uses this past data to find patterns in the characteristics of defendants who failed to appear in court, such as their age, number of past offenses, and the criminal history of their peers, and tries to distinguish them from those who did appear for their trial. This highlights an assumption built into much of predictive AI: people with similar characteristics will behave similarly in the future.

Predictive AI is quickly gaining in popularity. Hospitals, employers, insurance providers, and many other types of organizations use it. A major selling point is that it allows them to reuse existing datasets that have already been collected for other purposes, such as for bureaucratic reasons and record keeping, to make automated decisions.

Yet, predicting the future is hard. People could face unexpected setbacks such as getting evicted, or surprises such as winning a lottery, that no model can predict. Small changes in someone's life, such as a visit to the emergency room, could have large effects on their future—say, due to compounding medical bills.

Claims about the virtues of predictive AI are pervasive. Upstart's model decides whether an applicant should be approved for a loan.[13] The company claims that its model is significantly more accurate than traditional lending models. It also claims to be a leader in fair lending practices and promises that future versions of the model will continue to be fair. Finally, it claims to be highly efficient: three-quarters of its loan decisions require no input from humans. Other companies make similar promises. HireVue sells tools to automate hiring decisions. It claims to predict how well a candidate will perform at their job once hired. HireVue's home page announces its selling points: "Fast. Fair. Flexible. Finally, hiring technology that works how you want it to."

Despite these statements, human decisions still come in at many stages of the development of predictive AI, often hidden from sight. And since the data used to train predictive AI is itself created using human discretion, we cannot guarantee that the decisions will be unbiased or fair.[14] In other words, decisions made using predictive AI may still be very human.

We were skeptical of the promises made by developers of predictive AI, and so we decided to dig deeper. Together with researchers Angelina Wang and Solon Barocas, we spent over a year reading hundreds of research papers, news articles, and reports about automated decision-making systems. We were surprised to find that the same set of shortcomings plague many applications of predictive AI.[15] In the next few sections, we will take a closer look at these shortcomings, through real-world case studies.

Spoiler: We think predictive AI falls far short of the claims made by its developers.

A Good Prediction Is Not a Good Decision

Healthcare workers make a critical decision when a patient comes to the hospital with symptoms of pneumonia: whether to send the patient home after treatment or admit them overnight. They consider a patient's age as well as preexisting medical conditions, such as asthma, which place them at high risk if they get pneumonia. High-risk patients with pneumonia symptoms are typically sent straight to the ICU to avoid complications.[16]

In a 1997 study, researchers investigated whether AI could make better decisions than healthcare workers in predicting outcomes of patients with pneumonia.[17] Like countless other AI researchers, they thought a model trained with a sufficiently large amount of data would make better decisions than a human, decisions which would help to prioritize high-risk patients.

The researchers trained an AI model and found that it had fairly good accuracy in predicting which pneumonia patients would face complications or death. Surprisingly, the model found that having asthma leads to *lower* risk of complications due to pneumonia. If this model were to be used in a hospital, it would be more likely to send an asthmatic patient home than a non-asthmatic one (let alone sending asthmatic patients to the ICU). How is this possible?

The researchers looked more closely. They found that in the dataset, asthmatic patients were indeed at a lower risk for serious pneumonia or death—but this was only because the training data was collected under the hospital's preexisting decision-making system.[15] Asthmatic patients were sent straight to the ICU as soon as they arrived, thereby receiving more intensive care than non-asthmatic patients and thus becoming less likely to develop complications.

So the model's predictions were correct—but they were correct under the hospital's existing system. Ironically, the model was meant to replace this system.

Deploying this model would have led to disastrous decisions. Asthmatic patients would have been perceived as lower risk by the model and sent home. Thankfully, the researchers realized what had happened and did not use the model in this particular hospital.

This error shines a light on a fundamental limitation of much of predictive AI: AI can make good predictions *if nothing else changes*. But correlation is not causation (and having asthma does not reduce the risk of complications from pneumonia). In other words, predictive AI does not account for the impact of its own decisions. That is, it can't tell us what would happen if something changed in a system—if, for instance, the model started sending asthma patients home.

Let's look at another example from healthcare. A 2018 study claimed to accurately predict hypertension using machine learning, with impressive results.[18] On closer look, however, it became clear that the model was evaluated on people who had already been seen by clinicians.[19] This meant that a critical input to the model was whether a patient already used drugs for controlling hypertension. But if a patient is using antihypertensive drugs, they have obviously already been diagnosed with hypertension. These cases were still counted as successes when evaluating the model, vastly inflating its accuracy numbers.

This issue arose in part because the researchers relied on existing data rather than new data collected specifically for the task. Generally, collecting data is expensive and time-consuming. So, some AI developers claim that existing data is enough to make good decisions.

The importance of collecting new data is widely recognized in medicine, which relies on randomized controlled trials (RCTs) to test the impact of introducing a new drug or vaccine. To test if a drug performs well, half the people in the trial are randomly assigned to the experimental group and are given the drug. The other half are assigned to the "control" group and are given a placebo. The groups are carefully balanced with respect to demographics like age and gender. A drug's effectiveness is evaluated based on whether it leads to lower disease rates in the experimental group compared to the control. Medical researchers perform RCTs despite their slow pace and high expense for a simple reason—easier, faster methods don't work. The same is true in many areas where AI is used for automated decision-making.

Unfortunately, predictive AI companies don't seem to have realized the importance of collecting data on the impact of their tools' decisions. And such data collection is also expensive and time-consuming, so it would negate their claims of cutting costs and increasing efficiency.

As a result, even if AI can make accurate *predictions* based on past data, we can't know how good the resulting *decisions* will be before AI is deployed on a new dataset or in a new setting. So, when you encounter claims about the effectiveness of predictive AI, it's important to find out if developers evaluated the impact of the decisions or only the accuracy on past data.

Opaque AI Incentivizes Gaming

We've seen that we can't fully determine the impact of AI before deployment, in part because AI makes its predictions assuming that the system will remain the same as it was during testing. But systems change, and people are one obvious wild card. Things become even more complicated when people behave strategically.

The colonial British government in India wanted to reduce the cobra population, so it decided to offer a reward to people who brought in dead cobras. But instead of killing cobras in the wild, people started breeding them to claim the reward—leading to an *increase* in the cobra population. This is an apocryphal example of a mismatch between what is specified and what is desirable. A similar problem occurs in AI. When developers build AI, they specify the exact outcome they want to predict. But because of the reliance on past data, they are forced to choose this outcome based on the data that is already available. As a result, what AI predicts can be a poor proxy for what we want it to predict.

A stark example is AI used for hiring. In the United States, three-fourths of employers use automated tools to screen job candidates.[20] Automated hiring tools come in various forms. Some are used to filter candidates based on their résumés. Others test candidates using an automated video interview. Yet others require them to solve puzzles. These tools are used as the first step in selecting candidates. If someone doesn't pass, their résumé is discarded without a human ever looking at it. But the process is opaque; companies don't share how their software is built, and candidates are left in the dark about what criteria they are judged on.[21]

In response, candidates have developed strategies to work around opaque hiring AI. They stuff their résumés with keywords from the job application and add the names of top universities in white text (which a human reader can't see, but a computer can recognize).[22] In video interviews that they know will be judged by AI, they try using fancy words such as "conglomerate."[23]

It is not clear whether these strategies work. So a group of journalists decided to investigate. They looked at Retorio, a Munich-based startup that offers hiring tools based on video interviews.[24] Their findings were surprising: simple changes to a person's appearance, such as wearing a scarf or glasses, were enough to drastically change the score that the AI tool produced. Adding a bookshelf or a painting in the background increased scores, while making a video darker (without changing its content) led to lower scores. In another study, researchers looked at personality tests used in hiring and found that merely changing the format of candidates' résumé from PDF to plain text changed their personality scores.[25]

Changing the background or format of a résumé obviously does not change someone's capability to perform well at a job. So why do candidates' scores change? One reason could be that in the data used to train the model, people who had bookshelves in the background performed better compared to those who had a plain background.

This is precisely the kind of correlation that candidates banked on when they added keywords to their résumé. They assumed people with those keywords in their résumé performed better in the past. They wanted to avoid getting rejected just for omitting the right keywords or not using enough fancy words.

Acting strategically—gaming—to avoid negative consequences is common. This is what teachers do when they teach to the test, and what consumers do when they try to increase their credit scores without changing their spending habits, such as by getting a retail credit card or filling out a prequalification form before applying for credit. But with AI-based hiring, people don't even know what actions influence their chances. Instead of guiding candidates toward decisions that actually improve their skills, AI-based hiring encourages candidates to experiment with superficial changes to their résumé and application.

We are not taking a stance on whether candidates should or shouldn't be attempting these tricks. The ethics of gaming are interesting but tangential to our point. Rather, our point is that when AI companies make claims about accuracy, they fail to account for the effects of people behaving strategically. When the model's output can be easily manipulated using superficial changes, we cannot take claims of accuracy at face value. Opaque models also impose a cost on decision subjects in terms of wasted time, such as when candidates experiment with ways to add invisible fake qualifications to their résumés.

Overautomation

In 2013, the Netherlands deployed an algorithm to flag welfare fraud, replacing an earlier system in which humans reviewed each decision.[26] The algorithm was used to make serious accusations of guilt using only statistical correlations in data, without any other evidence.[27]

This shift to an automated system had many adverse effects. For one, people lost the ability to challenge decisions. Inaccurate or outdated government data often leads to erroneous fraud allegations, which under the new system could not be reversed. Further, the data used to make these allegations was not publicly available, so people had no way to find out why they'd been accused of fraud.

Over the next few years, the algorithm wrongly accused around thirty thousand parents of welfare fraud in childcare payouts. The amounts the government said they owed were in some cases over a hundred thousand euros—sending many parents into mental and financial ruin.[28] Shockingly, the algorithm used nationality to predict if someone had committed fraud. All else being equal, people were more likely to be flagged for fraud if they had Turkish, Moroccan, or Eastern European nationality.[29]

Despite the algorithm's shortcomings, the country used it for six years. When details of the algorithm were shared in 2019, the public was outraged. The Dutch data protection watchdog investigated privacy failures caused by the algorithm's use. It fined the country's tax authorities—who had built the algorithm—EUR 3.7 million. This was the largest such fine in the country. In 2021, the prime minister and his entire cabinet resigned, in part due to the use of the welfare fraud algorithm.

This is an example of overautomation: when AI is used to make decisions without offering any recourse to decision subjects. Overautomation has led to other notable failures in fraud detection—even when AI is not in use. Using an algorithm to detect unemployment fraud, the U.S. state of Michigan wrongly collected USD 21 million from residents between 2013 and 2015.[30] And the Australian government incorrectly collected AUD 721 million from its citizens between 2016 and 2020 in what's been called the Robodebt scandal.[31]

To protect themselves from accusations of overautomation, developers of these systems often include fine print saying that they should always be used with human supervision. But this merely passes the buck and doesn't necessarily have the intended effect.

In the summer of 2022, Toronto used AI to predict when people should avoid swimming at public beaches due to high bacteria levels.[32] The developer claimed the software achieved over 90 percent accuracy in predicting water safety. But it did much worse: on 64 percent of the days when the water was unsafe, beaches remained open based on incorrect assessments.

When journalists questioned city officials about the efficacy of the predictive tool, officials responded by saying it wasn't used on its own—a human overseer always made the final decision. Journalists later found that the supposed overseers *never* changed the decisions that the software spit out.

This is a familiar pattern. AI developers use a bait and switch when it comes to human oversight. They sell predictive AI based on the promise of full automation. Eliminating jobs and saving money is a big part of their pitch. But when AI fails, developers retreat to the fine print, saying that it shouldn't be used without human oversight.

Even if that oversight technically exists, it is often inadequate due to limitations of time, expertise, or authority. The bureaucrats in charge may be overworked, or may not have the training to challenge the automated decisions or the incentive to stick their necks out to do so.

In one extreme case, U.S. health insurance company United-Health forced employees to agree with AI decisions even when the decisions were incorrect, under the threat of being fired if they disagreed with the AI too many times. It was later found that over 90 percent of the decisions made by AI were incorrect.[33]

Even without such organizational failure, overreliance on automated decisions (also known as "automation bias") is pervasive. It affects people across industries, from airplane pilots to doctors. In a simulation, when airline pilots received an incorrect engine failure warning from an automated system, 75 percent of them followed the advice and shut down the wrong engine. In contrast, only 25 percent of pilots using a paper checklist made the same mistake. If pilots can do this when their own lives are at stake, so can bureaucrats.[34]

No matter the cause, the end result is the same: consequential decisions about people's lives are made using AI, and there is little or no recourse for flawed decisions.

Predictions about the Wrong People

AI reflects its training data. It learns patterns about the people who make up the data, and the decisions made by AI reflect these patterns. But when the decision subjects come from a population with different characteristics than those in the training data, the model's decisions are likely to be wrong. For instance, a predictive AI system that performs well in one country could be utterly useless in another.

Let's look at how this plays out in two criminal risk prediction systems in the United States: the Ohio Risk Assessment System (ORAS) and the Public Safety Assessment (PSA). Similar to COMPAS, both attempt to predict the risk of releasing a defendant before trial.

ORAS was trained on data from 452 defendants from the U.S. state of Ohio in 2010 and is now used across the United States. There are a few obvious problems with this. Ohio's patterns of criminal activity might differ from those in other states. The data doesn't represent everyone. The small number of people who were used to create the model could otherwise vary from the larger population it was eventually used on. Finally, as with any model, accuracy could degrade as patterns of crime change over time.[35]

By contrast, PSA was trained using 1.5 million people from three hundred jurisdictions across the United States. It is used in over twenty states. At first glance, this seems to solve part of the problem. If the model is trained on a much larger dataset from many jurisdictions, it should be more accurate when applied to the entire country. But nationwide trends of criminal activity can differ from local trends.

Cook County in Illinois is a stark example. It adopted PSA in 2015. But the rate of violent crimes there is much less than the national average. Compared to the training data, *ten times* fewer defendants who were marked as "high risk" went on to commit violent crimes. PSA used data from all over the country, but it did not consider that crime rates could be much lower in some counties. Thousands of defendants were unnecessarily jailed for months before trial—based only on a model's prediction, without any evidence of their guilt.[36]

PSA erred because it didn't distinguish data from different counties—it made predictions about the wrong people. In some cases, getting data about the entire population of interest is impossible, making this error hard to resolve.

One example comes from Allegheny County, Pennsylvania. In 2016, the county adopted the Allegheny Family Screening Tool to predict which children were at risk of maltreatment.[37] The tool is used to decide which families should be investigated by social workers. Through these investigations, social workers can forcibly remove children from their families and place them in foster care.

The tool relies on public welfare data, which consists mainly of data on poorer parents who use public services such as Medicaid-funded clinics. Notably, it doesn't include information about people who use private insurance.[38] Models built using this data therefore cannot make decisions about rich parents who have never relied on public services. As a result, the tool disproportionately targets poorer families.

This is an example of how AI tools search under the streetlight. More often than not, the streetlight is pointed at the poor.

Whenever predictive AI is deployed, it is critical to ask: Who was it tested on? When predictive AI is built using one population but is used on another, claims about how well it performs are based on insufficient evidence.

Predictive AI Exacerbates Existing Inequalities

The cost of flawed AI is not borne equally by all. The use of predictive AI disproportionately harms groups that have been systematically excluded and disadvantaged in the past.

A notable example comes from predicting who should receive better medical care. With the passage of the Affordable Care Act in the United States in 2010, insurance companies started asking hospitals to provide services at lower prices, threatening to remove providers who did not comply. One of the main ways hospitals cut costs was by identifying high-risk patients and providing them with preemptive care so as to avoid future expenses, such as hospitalization. To identify high-risk patients, they turned to AI.

Dozens of models were developed to assess patients' health risks. Developers claimed that AI could rank patients based on their healthcare needs and assign additional resources to the ones classified as high risk. While the pneumonia risk prediction model that we saw in the beginning of this chapter was never deployed, health risk prediction tools used to reduce hospital expenses are deployed across the United States.

Enrollment in a high-risk healthcare program has huge consequences. It dictates whether a patient will receive preemptive care and personal assistance. Still, most developers did not share how they built their models—until recently, we knew little about how well their products performed.

One such product is Optum's Impact Pro. In 2019, researchers studied it and found that based on its predictions, Black people were less likely to be admitted to high-risk programs than White people. In other words, a Black person who had the same health risks as a White person would receive poorer care.[39]

On closer inspection, the researchers found the culprit. It is hard to quantify how much a patient needs healthcare. But hospitals know exactly how much people *spend* on healthcare. So instead of predicting healthcare *needs*, Optum chose to predict how much an insurance company will *spend* on healthcare.

Higher healthcare costs don't always reflect higher health needs. Higher bills could occur due to access to better insurance; more time, attention, and care at the doctor's office; or more visits to the doctor. Or perhaps those with higher bills are simply those who can afford to pay higher costs for co-pays and deductibles. There are clear inequalities in the U.S. healthcare system, and the use of Optum's predictive AI would mean that people who were already receiving better healthcare would be more likely to be classified as high risk and continue to receive better care in the future.

Unsurprisingly, in the training data used for creating the model, Black people received poorer care than White people, even those with similar health conditions. The tool had a racial bias because the predicted quantity (healthcare costs) was a poor stand-in for what the developers claimed to predict (healthcare needs or a patient's risk level).

Optum's use of healthcare costs to measure risk made sense for its business. Its clients were hospitals that wanted to reduce costs. Even after this damning study was published, the company continued using healthcare costs to build its model.

Business incentives are one of many reasons predictive AI increases inequality. Another is developers' reliance on past data. As we have seen, collecting new data for developing predictive AI is expensive and time-consuming. But existing data might not have information about what developers want to predict (in this case, patients' healthcare needs or their risk levels). So, developers use proxies that are easier to measure and already exist in the data (such as healthcare costs).

Let's look back at COMPAS, the tool which provides a pretrial risk score. Its developers claim that judges can use this score to predict whether a defendant will commit a crime or fail to show up for their trial. But the data used by COMPAS does not contain information about crimes. It only includes information on who was arrested. The difference is subtle but salient. Not all crimes lead to arrests—in many cases, they can be undetected or ignored. Police could arrest someone only to later find them innocent in court. And there are well-known racial disparities in policing—Black people in the United States are more likely to be arrested than White people for the same crimes. These differences can create chasms between what companies claim to predict (crime) and what they actually predict (arrests).

For all these reasons, when predictive AI systems are deployed, the first people they harm are often minorities and those already in poverty. We've already seen other examples, such as Allegheny County's child maltreatment prediction tool that only had data about those who access public benefits such as child welfare. And we'll see many more in the chapters to come.

A World without Prediction

Why is predictive logic so pervasive in our world? We think a major reason is our deep discomfort with randomness. Many experiments in psychology show that we see patterns where none exist, and we even think we have control over things that are, in fact, random.[40,41] When people are forced to confront randomness and the illusion of control breaks down, they look for ways to take back control.

A good example is election forecasting. In the United States, presidential elections occur every four years, and forecasting the winner is a common practice, starting well over a year in advance. But it is far from an exact science. On the morning after the 1948 presidential election, the *Chicago Tribune* infamously blared "Dewey Defeats Truman" on its front page.*[42] Of course, Truman had won the election and would be the next U.S. president. In a hurry to call the election before the paper went to press, the newspaper relied on poll outcomes to preemptively predict the (incorrect) winning candidate. And seventy-five years later, things are only slightly better. The 2016 election forecasts were famously incorrect, predicting Hillary Clinton defeating Donald Trump.

Despite its limitations, forecasting has become a sort of spectator sport, and people's fascination with it can border on obsession. Here's a quote from a *Wired* article about one such person:

> *When he wakes up in the morning, he doesn't shower or eat breakfast before checking the Nate Silver-founded site's presi-*

---

* Harry S. Truman and Thomas E. Dewey were the Democratic and Republican presidential candidates in the 1948 elections.

> dential election forecast. He keeps a tab open to FiveThirtyEight's latest poll list; a new poll means new odds in the forecast. He gets push alerts on his phone when the forecast changes. He follows the 538 Forecast Bot, a Twitter account that tweets every time the forecast changes. In all, Evan says he checks in hourly, at least while he's awake.[43]

This is despite poll numbers not affecting most people's daily lives in any meaningful way. Why does this happen? It all comes down to our inability to handle the uncertainty hanging over our lives. Uncertainty avoidance goes one level deeper than being averse to risk. Even if the election forecaster says the chance of our favorite candidate winning is 50-50, that somehow feels better than not knowing at all.

What does this have to do with AI snake oil? Here's our thesis. It's true that companies and governments have many misguided commercial or bureaucratic reasons for deploying faulty predictive AI. But part of the reason surely is that decision-makers are people—people who dread randomness like everyone else. This means they can't stand the thought of the alternative to this way of decision-making—that is, acknowledging that the future cannot be predicted. They would have to accept that they have no control over, say, picking good job performers, and that it's not possible to do better than a process that is mostly random.

And what if a decision-maker did embrace unpredictability? Suppose a company announced that it would hire candidates randomly after filtering out clearly unqualified candidates, and it would promote employees randomly among those who met performance criteria. Because the overemphasis on merit is so ingrained in so many people, the company would come to be seen as an undesirable place to work, and it would lose good candidates. Such a policy wouldn't survive long in our world.

In fact, when such lotteries were used for allocating housing, people participating in those lotteries had a largely unfavorable opinion of them.[44]

And so, the discomfort we experience with randomness can lead to a search for patterns where none exist. This paves the way for biases. Consider the business mantra, "No one ever got fired for buying IBM." Opting for the familiar, the tried-and-true, is safer than embracing uncertainty. Even the practice of hiring primarily from elite colleges reflects a desire to impose order on the inherently unpredictable process of identifying talent and potential.

Accepting the inherent randomness and uncertainty in many of these outcomes could lead to better decisions, and ultimately, better institutions. Instead of treating people as fixed and their outcomes as predetermined, we need to work on building institutions that are genuinely open to the fact that the past doesn't predict the future. Such a world is possible—if only we can learn to embrace the randomness that underpins our lives. We'll return to this theme in the final chapter.

Concluding Thoughts

As predictive AI is deployed, it is important to understand how it fails and how it harms people, in order to avoid falling prey to snake oil. Change starts by challenging the deployment of harmful AI tools in your workplace, neighborhood, and community. An informed public is the first step in demanding change.

In this chapter we encountered many reasons why predictive AI fails. Table 2.1 recaps them. But given the increasing amount

TABLE 2.1. Five reasons predictive AI fails

Reason	Example
A good prediction can result in a bad decision.	Patients with asthma could be sent back home when they come to a hospital with symptoms of pneumonia.
People can strategically game opaque AI.	Adding bookshelves in the background increases scores on automated hiring tools.
Users over-rely on AI without adequate oversight or recourse.	The Dutch welfare fraud detection model falsely accused 30,000 parents of fraud without any recourse.
Data for training AI may come from a different population than the one it is used on.	PSA's criminal risk prediction relied on a national sample. It overestimated the risk in counties where crime was rarer.
Predictive AI can increase inequality.	Optum's Impact Pro led to an increase in the difference in the quality of care between Black and White patients.

of data collected about people, as well as advances in machine learning, it might seem like the limitations we have seen are temporary. On the other hand, it is also possible that no matter how much data we have or how good our models become, there are inherent limits to how predictable the future is. Which of these scenarios is more likely? The next chapter will answer this question.

Chapter 3

WHY CAN'T AI PREDICT THE FUTURE?

ARE OUR LIVES PREDETERMINED? Can we predict what will happen to us tomorrow? Next year? In our careers? And if it is possible to predict our futures, how do we go about doing it? These questions are not new. Humans have always wanted to know the future. Historically they've relied on the predictions of individuals who claimed to be "gifted" in discerning the future. Kings consulted oracles before they went to war. People went to fortune tellers in an unpredictable world where diseases could strike at any moment.

Rather than relying solely on the purported clairvoyance of certain individuals, people throughout history also developed prediction systems that anyone could conceivably master.[1] These systems were structured ways of predicting the future. They were based on the positions of the stars, or tarot cards, or lines on the palm, or dozens of other things.

Today, the prediction system of choice is AI. Many people believe that we can use algorithms and machine learning to analyze current and historical facts in order to make accurate predictions about future events. An apocryphal example of gaining surprising insights from data is an unlikely story of beer and diaper sales.[2] When analyzing sales data from a chain of stores, a data scientist learned that the sales of beer and diapers were correlated—presumably because new parents don't have time to go to a bar for a drink, but they can drink at home. So, if stores stocked these two items next to each other, the sales of both items would increase because when parents were shopping for diapers, they would also pick up beer (and vice versa).

This type of insight seems like the perfect example of what we can understand from data. After all, there could be millions of such correlations possible in a typical store, and it is believable that AI could indeed tell us which of these correlations we should take seriously. Similarly, techno-optimists have argued that statistical tools can replace the scientific method because of the sheer number of correlations they can find in data that no human scientist could conceive of.[3]

It's understandable why this is a popular idea—why we might believe that computers, through sheer number crunching, can foretell what's to come. Unlike other methods of predicting the future, the power of computers doesn't require a suspension of disbelief—after all, computing, and AI specifically, has worked extremely well in many areas, such as transcribing speech or creating images from text. So, to the average person, it doesn't feel like too much of a stretch that a company could develop a tool to predict future outcomes, say, one that tells police where and when crime will occur.

At the same time, in the last chapter, we saw that predictive AI often doesn't work as claimed, and it can cause a lot of harm. Are these limits inherent to predictive AI? Or will current limits be overcome as AI methods advance and we gain access to more data?

Given the rapid uptake of predictive AI, in 2020, Arvind co-taught a course with Matt Salganik at Princeton University that collated evidence on how well we can predict the future.[4] This chapter is based on the lessons from that course and the research Sayash and Arvind have done since then. The main questions that we will address are: How well can AI predict individual and societal outcomes, and where does it fall short? In which cases can we hope to improve our predictions about the future, and in which cases are limits to predictability inherent?

Let's begin by looking at the computing tools we use to predict the future.

A Brief History of Predicting the Future Using Computers

When your weather app tells you there's an 80 percent chance of rain, you're witnessing the end result of a millennia-long quest to understand the skies. In Egypt, before the construction of the Aswan High Dam, the Nile flooded every year. Starting in the third century BC, and continuing for about a thousand years, priests observed the rising level of the Nile to predict water levels for each farming season.[5] If the level reached a stone marker by a certain date, priests would know the flooding that year would be adequate for farming.

Fast-forward a thousand years. By the end of the seventeenth century, several meteorological instruments had been created: hygrometers to measure humidity, thermometers to measure temperature, and barometers to measure pressure. In the nineteenth century, weather-observing stations were set up across the globe. Compiling data from different locations around the world could enable meteorologists to predict, to some extent, how the weather would evolve. In the early twentieth century, equations to simulate weather patterns were proposed, but it wasn't until the 1940s that the computational power required for these calculations caught up. John von Neumann, one of the most prominent mathematicians in the world at the time, directed the construction of a computer at the Institute of Advanced Study in Princeton (less than two miles from where we are writing this book). Among other applications, it could calculate weather prediction estimates using the simulation equations that had already been proposed. By the mid-1950s, regular weather forecasts were being made across North America.[6]

But a 1963 finding raised doubts about our ability to predict the weather well—especially for longer time horizons.[7] Meteorologist Edward Lorenz was trying to predict weather patterns using simulation equations. During one of these calculations, he made a chance discovery: rounding the numbers to three decimal places instead of six gave vastly different results![8]

This finding led to a profound scientific advance—the recognition that weather is a chaotic system. That is, small changes in initial conditions, like a small error in measuring temperature, lead to exponentially increasing errors later. The farther away the prediction, the larger the error. Lorenz termed this the Butterfly Effect: the flapping of a butterfly's wings in Brazil can cause a tornado in Texas.[9] What this means is that at least in principle, a butterfly flapping its wings could have ripple effects on the atmosphere that grow larger and larger with time. Once scientists understood this effect, predicting the weather over longer time periods started to seem like a herculean task.

But something interesting has happened in the last few decades. Increased computational power, more data, and better equations for simulating the weather have led to weather forecasting accuracy increasing by roughly one day per decade. A five-day weather forecast a decade ago is about as accurate as a six-day weather forecast today.[10] Of course, it remains impossible to predict the weather a year out or even a month out. To do that, you'd really need to know the position of every butterfly. But a one-week forecast is still extremely valuable. This improvement is not the result of a revolution in computational methods, theory, or data collection capabilities. Rather, it is the end result of consistent small improvements.

Weather prediction today largely relies on a computational tool known as simulation. The idea is that the future evolution of a system can be predicted using two crucial pieces of information: the current state and equations describing how the system changes over time based on the interactions between its components.

Given the successes of predicting the evolution of physical systems, one might hypothesize that with the right data and enough computing power, simulations could be used to predict any type of event. In the 1950s, MIT professor Jay Forrester pioneered an approach called system dynamics, which tried to extend the use of simulations to social systems.[11] He tried to use simulation to model a whole city, with an eye to solving thorny social problems like urban blight.

Here, things went badly wrong. And we don't have to speculate too much to find out why—a city is far from a physical system. A city can't be perfectly observed, even in a totalitarian society, so there is a lot of uncertainty in the starting conditions of a simulation. Further, a city is not a self-contained system, so unless you're simulating the whole world, your simulation will go off track. In fact, Forrester's models didn't account for suburbanization at all—the growth of the suburbs as residential areas, with people commuting into cities for work. This was perhaps the most significant shift that American cities were undergoing at the time of his work.

In the '60s, a company called Simulmatics—a portmanteau of *simulation* and *automatic*—aimed to predict outcomes like U.S. elections, the effectiveness of counterinsurgency efforts in Vietnam, and race riots.[12] As far as we can tell, it didn't actually use simulation at all, just basic statistics and algebra.[13] Still, the company's choice of its name was perhaps indicative of the extent to which people at the time associated simulation with prediction. But Simulmatics, too, was largely a failure, and the company shut down in 1970.

In contrast to simulation, machine learning uses past data to learn underlying patterns and make predictions about future events, often in ways that can adapt and change over time. There are no fixed rules about how the future will play out given past events. Rather, these rules are determined based on how the system behaved in the past.

For example, weather simulations typically rely on physical laws about the world, such as equations describing airflow over time. But if we were to use machine learning, we would use data to build computer models of how the weather has evolved in the past, and then use those models to predict how it will evolve in the future. We would no longer need to rely on an understanding of physics. Nor would we need to predict the weather minute by minute or hour by hour, with each prediction being the input to the next. We would instead directly model the relationship between the weather today and the weather tomorrow, or whatever time horizon we're interested in.

Broadly speaking, machine learning is more suited for predicting things about individuals whereas simulation is more suited for predicting collective or global outcomes. That's because there are millions of people about whom you can collect training data to build machine learning models. For example, spam classifiers today work reasonably well, because there are lots of examples of both spam and nonspam emails.

But if you're predicting whether there will be a food shortage, machine-learning-based predictions are not that useful, because there aren't enough examples of food shortages in the past that can be used to train a machine learning model. Here, simulation based on domain knowledge, such as how changing climate conditions affect agriculture, the intricacies of global trade, and the impact of political shifts on food supply can be more useful.

Around the same time as the fall of Simulmatics, the seeds for more sophisticated machine learning methods were taking root. FICO* was established in 1956 to predict individuals' credit risk. The company started by creating models for individual lenders and banks. Much later, in 1989, the company launched the general-purpose FICO score, which could be used as a measure of creditworthiness across financial institutions. The FICO score used the credit history of individuals to predict how likely they would be to default on a loan in the future.[14] Similarly, in the 1980s, criminal risk prediction became widespread.[15] Soon after the turn of the twenty-first century, machine learning became commercially successful in serving personalized ads, and then in many other areas. It started to be seen in the tech and business worlds as the default approach to almost any decision problem. Following this success, the use of predictive AI exploded in the 2010s. The vast majority of large companies started using some form of hiring automation.[16] And we'll see many more examples soon.

---

* Then known as Fair, Isaac, and Company.

Despite the widespread use of computational predictions in social settings, you might be surprised to learn that there's remarkably little published evidence of their effectiveness. So, how well can we really predict future social outcomes? Let's dive in.

Getting Specific

Not all predictions are hard. Some phenomena can be predicted with a high level of accuracy. Astronomers can predict planetary movements and the life cycles of stars. Doctors can predict the progression of some diseases—and they can even predict how different drugs affect that progression. Given these successes, what are phenomena that we *can't* predict?

Based on what we saw in the previous chapter, a first guess could be that we can't predict any social phenomena. But even this is too broad. We can certainly predict some social phenomena reasonably well, like the amount of traffic on a route, or how busy a store would be on a certain day.[17,18] So, clearly, we need to be more specific.

One area where predictions seem hard is people's futures. Can we predict a student's GPA a year in advance? What about predicting if someone will be evicted from their current home? In the next section, we'll review evidence of how accurate such predictions can be. But first, let's pause to ask how accurate a prediction needs to be in order to be considered good or bad. This question might seem simple, but it turns out to be surprisingly nontrivial.

For virtually any kind of prediction, there are many different ways to judge its accuracy, ways which may produce results that seem drastically different from each other. When trying to predict the weather, do we consider a prediction accurate if the temperature is within one degree? Or is it enough to predict it within five degrees? Or do we only care if we can correctly predict if it will rain, regardless of temperature? Any of these criteria can help us compare two different methods for prediction or determine whether our ability to predict the weather is improving over time. But there isn't one true way to judge if a weather prediction is good or bad.

This also means we can't directly compare how good a prediction in one area is to a prediction in another area. That is, we can't make statements like "weather patterns are harder to predict than sales patterns." It isn't meaningful to ask whether predicting the temperature within one degree is better or worse than predicting the sales amount in a store within 10 percent.

Still, there are many *qualitative* criteria that can help us understand whether prediction tasks can be done well. Weather forecasting isn't perfect, but it can be done well enough that many people look at the forecast in their city every morning to decide whether they need an umbrella. But we can't predict if any one person will be involved in a traffic accident on their way to work, so people don't consult an accident forecast every morning.

This comparison highlights another important quality of predictions: we only care about how good a prediction is in relation to what can be done using that prediction. For example, our ability to predict earthquakes is excellent—if the goal is to identify high-risk zones that need to have stringent building codes.[19] But if the goal is to identify when an evacuation needs to take place because of an imminent earthquake, our prediction ability is basically nonexistent.[20] The first task is about where earthquakes might strike, while the second task is about when.

In other cases, such as in criminal risk assessment, prediction could be practically useful from the perspective of those in power—say a court system trying to manage jail populations—yet morally dubious in deciding who should be jailed and who should be released.

A third, related criterion is whether the accuracy of the prediction improves with more data and better models. Suppose a disease is the result of a complex but deterministic genetic process. That is, it depends completely on someone's genes. Then, with enough genetic sequencing and improved models, we can hope to predict whom it will affect with high accuracy. On the other hand, suppose a type of cancer is the result of radiation exposure leading to a random genetic mutation. We might be able to say that someone has an elevated risk because they are exposed to radiation in their job, but beyond that risk assessment, we can't predict if they will develop cancer, no matter how much data we collect, because of the randomness inherent to this process.

For both diseases, the judgment of high and low predictability is not based on absolute numbers. In fact, in numerical terms, given our current (relatively poor) knowledge of genetics, genetic disease may be far less predictable than cancer. Yet the two trajectories of predictability differ as our data and scientific knowledge increase. Genetic diseases become more predictable over time, whereas there are fundamental limits to the predictability of cancer, just like we can't predict the outcome of the roll of a fair dice, no matter how much data we collect or how good our AI is.

So when we say life outcomes are hard to predict, we are using a combination of these three criteria: real-world utility, moral legitimacy, and irreducible error, that is, error that won't go away with more data and better computational methods.

Let's return to the question of why life outcomes are hard to predict. One of the most obvious explanations is that people have agency. This explanation is unsatisfying to us. Of course, people have agency, but how often do we exercise it? Maybe we just don't exercise it often enough, and as a result we are still largely predictable. Besides, even if people aren't inherently predictable in a state of nature, the kinds of things that organizations want to predict, like students' GPAs, are constructs that are specifically designed and chosen for the purpose of exerting control over individuals. So, they may be much more predictable than other aspects of people's behavior.[21]

To arrive at better answers, let's look at empirical evidence from the real world.

The Fragile Families Challenge

Predicting the future is a focus in many scientific fields, but not in social science. Instead of predictions, the dominant approach in social science is to focus on improving our *understanding* of what causes phenomena of interest.[22] For example, sociologists generally do not aim to predict individuals' future income for targeted interventions. Rather, they aim to understand the causes of poverty so that interventions can be targeted toward alleviating poverty.

Similarly, while social science helps us understand the causes of incompatibility and divorce in marriages, it isn't used to predict whether a couple will file for divorce.[23] In the 1990s and early 2000s, sociologists did attempt this, and they generated a lot of excitement about predicting which couples would divorce. But these studies were plagued with faulty methods that led to vastly exaggerated claims of predictive accuracy.[24] Essentially, the models researchers used would perform well for people whose information was included in the dataset used to train the model, but not on couples whose information wasn't in the dataset. In other words, prediction only worked well in cases where the answer was already known!

One reason for the focus on understanding over prediction is the lack of available data. Until recently, sociological datasets were small. Machine learning works best when applied to large datasets, so sociologists have historically relied on simple statistical models like linear regression.

As the amount of available data has increased, the use of machine learning for prediction has made inroads in the social sciences. Let's take a look at one such attempt, the Fragile Families Challenge, a notable large-scale study that tried to predict children's outcomes using AI and lots of data.

In 2015, our colleague at Princeton University, Matt Salganik, wanted to study how well AI could predict the future. Sara McLanahan, also a Princeton sociology professor at the time, had been following the lives of over four thousand children who were born around the year 2000 in over twenty U.S. cities. Over the previous fifteen years, she and her colleagues had surveyed each child and their family—at birth, and at the ages of one, three, five, and nine. Through these surveys Sara and colleagues collected over ten thousand data points about each child using detailed surveys from parents, teachers, and in-home activities with the child. In fact, you'd have trouble finding a sociological variable of interest that *wasn't* included in the surveys.

In 2015, Sara and colleagues were planning to release their latest round of survey data, collected when the children reached the age of fifteen. Matt wondered if it was possible to use survey data from the Fragile Families project to see how well AI could predict the future. He walked down to Sara's office to discuss the details, and the seed of a collaboration was planted.

Matt and Sara felt that a single research group analyzing this data wouldn't suffice to evaluate whether children's life outcomes are predictable. It might give a lower bound on accuracy, but no matter how good the results were, someone might object that a different model could perform better. Or that a different group of better-skilled researchers would build better models. In other words, it is hard to prove a negative result in this setting.

To address this concern, Matt and Sara organized a prediction competition, together with Princeton colleagues Ian Lundberg and Alex Kindel. A subset of the data—all data from birth until age nine—was released to competition participants across the world. Using this data, participants were asked to create AI models that would predict how well the children were doing at age fifteen. They would make predictions about six outcomes for children, such as their GPA, whether they had been evicted, and whether their household was facing material hardship. Participants would be ranked based on how close they were to the true outcomes of the children.

Since the competition was open to the public (and hundreds of researchers participated), different teams could try out different methods. Some used complex AI models. Others used sociologically informed statistical models. Regardless of their approach, researchers would compete on an equal footing: the only thing that mattered was how well their model predicted future outcomes of the children. The goal wasn't to pick a winner, but to learn from the collective effort. In fact, the organizers called it a "mass collaboration."

But unlike a regular collaboration, the prediction competition would help avoid overoptimism about how well AI could work. Many AI studies exaggerate how well a model can predict the future, because the researchers building the model have access to the data used to evaluate it. This means they can overestimate how well the models would do in the real world (and this is exactly what happened in the divorce prediction studies we mentioned earlier). In contrast, the Fragile Families organizers didn't release the data used to evaluate the models, so none of the teams could game the results before submitting their predictions, even unintentionally.

Eventually, 160 teams submitted their predictions. Among the models being compared was a simple model relying on basic statistical techniques, which was intended to serve as a baseline for comparing more complex AI models. This baseline model consisted of just four features: three related to the mother of the child and one related to the outcome at age nine. For example, for predicting GPA at age fifteen, the model consisted of the mother's race, marital status, and educational level, and how well the student did in school at age nine.

To Matt's surprise (and disappointment), none of the models performed very well—the best models were only slightly better than a coin flip. And complex AI models showed no substantial improvement compared to the baseline model consisting of just four features.[25]

In other words, tens of thousands of data points about thousands of families, a hundred and sixty competing researchers, and the most advanced AI models were not much better at predicting the future than decades-old regression models that used well-established sociological theories. The data showed that past GPA, race, and social class are useful in predicting future GPA. Sociologists had long understood these trends, so they were not exactly headline news.

Why Did the Fragile Families Challenge End in Disappointment?

Before Matt presents the results from the Fragile Families Challenge in academic talks, he usually asks the audience to guess how the teams performed. Some of the most optimistic audience members are computer scientists and data scientists, who have seen how well AI can perform in other domains.

When confronted with the disappointing result, this is also the group with the most follow-up questions and ideas for improving predictions. One of the most common questions: Are samples from four thousand families enough? They often refer to another contest that set off the deep learning revolution in 2012. It was a contest known as the ImageNet Challenge that asked participants to identify the contents of images using AI, and it consisted of 1.2 *million* labeled images.[26] (We'll get back to ImageNet in more detail in the next chapter.)

One possibility for improving social predictions is exactly the sort of brute-force intervention that computer scientists suggested in this case—use samples from more people. The hypothesis is that greater amounts of data and more computational power would lead to drastically better accuracy and in turn a breakthrough in social prediction.

This is one of many reasons we can't conclude that the results of the Fragile Families Challenge indicate fundamental limits to prediction. We just do not know whether this hypothesis is correct. In scientific fields whose theories are more or less settled, such as planetary orbits in astronomy, there is high predictability. We can predict a planet's position years from now with astounding accuracy. In other cases, there are fundamental limits to predictability, and we know what those limits are. For instance, the laws of thermodynamics help us estimate the behavior of a gas like oxygen or nitrogen as a whole, but not the behavior of individual gas molecules.

But so far, we do not have theories about the predictability of social problems. We can't predict the future well, *and* we don't know what the fundamental limitations of our predictions are.

The predictability of life outcomes has been explored over and over in science fiction. The premise of the dystopian science fiction film and short story *Minority Report* is that would-be criminals can be apprehended based on predictions about crimes that they have not yet committed. The plot of *Person of Interest*, a TV drama, centers around AI that can predict crimes. The fundamental tension in these pieces is usually between predetermination and free will, yet they ignore one important source of irreducible error in predicting social outcomes: chance events.

One notable feature of tasks where AI seems to work well, such as classifying what's in an image, is that once you have an image, say an image of a cat, it is easy to say what's in it. In this case, the irreducible error is small; humans (and AI these days) can correctly classify most images. Chance plays little role in determining the correct answer.

How high is this irreducible error for social predictions? Our understanding of social science and our theories of predictability are not yet at the point where we have a concrete answer, but there are reasons to think that error is high, in part due to chance events. People can face shocks that are completely unpredictable and that have a large effect on their life trajectory. No model can predict if someone might win a lottery or get hit by a car, for instance.

But how often do these unpredictable events happen? It might be true that a butterfly could cause a tornado, but if this happens once every millennium, maybe we don't need to worry about it. Much more common than large shocks are small initial advantages that are compounded over time. For example, a small bias in annual performance reviews (say because you have an antagonistic boss) could lead to large effects over the course of your career (you may not be promoted as quickly as others).

The difficulty of measuring these small differences leads to higher irreducible errors in predictions.

Let's come back to how much more data we would need to predict future outcomes well. We know that the number of samples needed to create accurate models increases sharply as the samples become more noisy. And social datasets have a lot of noise. In addition, the patterns underlying social phenomena are not fixed. Unlike images of cats, they differ greatly based on different contexts, times, and locations. What identifies success at one place and time might be completely uninformative for predicting success at another.

This means that we might need a lot of data from different social contexts for AI to predict the future well. Simply using past data is not enough—just like it's not enough to use polling data from a previous election to predict if a Democrat or Republican would win the next U.S. presidential election.

This raises an interesting possibility. Perhaps collecting enough data to make accurate social predictions about people is not just impractical—it's impossible. Matt Salganik calls this the eight billion problem: What if we can't make accurate predictions because there aren't enough people on Earth to learn all the patterns that exist?[27]

Just as important as the number of samples is what's in those samples. In the Fragile Families Challenge, each child had around ten thousand sociologically relevant features recorded in the data. But you could imagine this, too, not being enough. Here's why.

After the prediction competition concluded, Matt and colleagues tried to find out why these models had performed so poorly.[28] So they decided to meet the families who had been the most unpredictable, to find out what caused the predictions about them to be so incorrect. In one interview, they found that a struggling child had suddenly improved their performance in school. The reason? A supportive neighbor who had counseled the child, provided help with homework, and fed them blueberries. But there was no question in the Fragile Families data that asked children if someone outside the family was feeding them blueberries (or, perhaps more impactfully, helping them with homework). Could this be an example of a missing feature that, if present in the data, would lead to accurate predictions—say, because it signifies the presence of a supportive adult in the child's life? How many such features are today's datasets missing?

One way to build a more expansive dataset that could be used for more accurate predictions is to rely on data collected by governments. For example, the Netherlands has compiled data about people's families, neighbors, classmates, households, and colleagues. This dataset is extensive; it has records on each person living in the country—a total of 17.2 million people. On average, each person is connected to eighty-two other people, for a total of 1.4 *billion* network relationship records.[29]

Such data is obviously much larger and more complete than the Fragile Families Challenge dataset. It is a realistic alternative that *could* be used to predict social outcomes. If it is indeed possible to use this data to predict future outcomes of interest, we'll find out soon enough. There are several efforts, including a prediction competition, that are currently underway to test this hypothesis.[30]

Another possibility for additional data collection is via tech companies. People already spend significant amounts of time on platforms owned by companies like Google and Meta. Could the data they provide to these companies have insights that wouldn't otherwise be available?

We can indeed speculate about this possibility, as have many pop culture explorations of the relationship between technology and society. But at the end of the day, making predictions about people's life outcomes is too risky, reputationally and legally, for tech companies. And in any case, their business metrics are not tied to predictions about people's future outcomes, but rather what content people will engage with today. So, it is unlikely we'll settle the question of the long-term predictive power of online data any time soon.

An even more grand (and dystopian) possibility is to collect extensive information on each person everywhere, to create überdatabases about people. In a world like this, everyone would be under constant surveillance so their every action is constantly monitored. The NSA and Big Tech companies already have a trove of data about people, but we're talking about something far more invasive—tracking every spoken word, every movement, every action, and perhaps even every electrical signal in their brain.[31] Would such a world enable better predictions? To what end? And at what cost to our privacy?

Predictions in Criminal Justice

Ultimately, there aren't many examples of social science research on predicting life outcomes. And the evidence we do have suggests that we're quite bad at it. But as we saw in the last chapter, that hasn't stopped companies and governments from deploying tools to predict the future and make decisions about people based on those predictions.

There are dozens of AI tools and products made by companies that claim to predict individuals' futures.[32] These predictions are used to make life-changing decisions about people in healthcare, insurance, banking, and criminal justice.

Through these deployments, we can try to assess how predictable life outcomes can be. While companies have resisted giving out information about how their predictive AI works, researchers, journalists, and advocates have still managed to finagle information using companies' sales materials, interviews with employees, and public records requests. So we can look at how predictive AI has fared in the real world, and analyze the limits to prediction in these domains.

For example, one selling point of tools like COMPAS is that they can reduce bias and increase accuracy in the criminal justice system.[33] Since AI doesn't have the same biases that humans do, the story goes, it could be used to make decision-making objective and fair.

In 2016, Julia Angwin and her colleagues at ProPublica set out to investigate these claims. They ran an in-depth study on the use of COMPAS in Florida's Broward County to assess how predictions about ten thousand people panned out.

What they found was a big racial disparity. Among people who would *not* go on to commit crimes, Black people were twice as likely to be misclassified as high risk compared to White people.[34] Specifically, 45 percent of the Black people who didn't go on to commit crimes were marked as high risk, compared to just 23 percent of White people.

This racial bias became the subject of many academic studies and press reporting on risk assessment tools. But the report also revealed that the tool wasn't very accurate to begin with; it had a relative accuracy of just 64 percent.[35] Relative accuracy is calculated by looking at pairs of people where one would go on to commit a crime and the other wouldn't, and calculating how often the former was ranked as higher risk than the latter. An accuracy of 50 percent can be achieved by generating risk scores randomly! So a questionnaire that uses 137 data points about a person is only slightly better than deciding a defendant's fate based on a coin flip.

From a defendant's perspective, even if they have no intention of committing a crime, they could be jailed simply because their answers were similar to those of others who had gone on to commit a crime in the past.

Even the abysmal accuracy of COMPAS is likely to be an overestimate. To some extent, COMPAS is just predicting the biases of policing. For example, in a hypothetical world where everyone has the same probability of committing a crime in the future, but some people are more likely to be arrested (say, because they live in overpoliced areas), COMPAS would still be good at predicting future outcomes, just because people's locations determine their likelihood of being arrested. We have evidence of such biases in policing today, so there's little reason to expect that COMPAS isn't just predicting these biases.[36]

Follow-up studies had results that were even more damning than ProPublica's original findings. One study showed that COMPAS was no more accurate than the judgments of nonexperts.[37] In fact, instead of using 137 data points, using just two data points about an individual—their age and their number of prior offenses—was as accurate as COMPAS. The lower the age and the higher the number of prior offenses, the more likely someone is to be rearrested.

There's cause for skepticism even in this simple rule. According to the data, younger defendants are riskier, but from a moral perspective, we might want to treat younger defendants more leniently because they are still developing neurologically, are more susceptible to peer pressure, and are more capable of long-term change.[38] If we take age out of the equation, we are left with just one feature: the number of past offenses.

So a simple rule—the higher the number of past offenses, the higher the risk—might be more morally legitimate than one that uses 137 data points about a person. And it has other positive side effects: the rule is easily understood, people can challenge incorrect decisions, and it doesn't transfer power to unaccountable third parties (unlike COMPAS, a black box tool developed by a for-profit company).

Failure Is Hard. What about Success?

To recap, the last few sections have shown that in many settings it is hard to predict who will experience negative outcomes. Perhaps this is because failure is less about people's intrinsic qualities and more about circumstance. Surely, predicting success is different? Isn't it merit that determines who is successful? Being the fruit of years of hard work, success must be more predictable than, say, a spur-of-the-moment crime.

Well, the evidence suggests that if anything, luck plays an even bigger role in success than in failure.

Let's start with a personal story. Today, Arvind has a successful academic career. He applied to nine prestigious graduate programs back in 2004. Then the rejections started pouring in. After he was rejected by eight schools, he emailed the ninth school to ask if they had any news. They said they had no record of his application.

He panicked, then reached out to a friend who was a graduate student at the school in question. His friend learned that the application had landed in some other pile by accident. Arvind's disbelief was surpassed only by his amazement when the school actually accepted him. If it weren't for the friend willing to locate his application, his graduate school aspirations would have ended in a wipeout.

Getting into graduate school was only one step. Looking back at his career, each rung of the academic ladder involved a dose of luck.

All successful people are lottery winners to some extent. But they rarely acknowledge the role of luck in their success. Arvind isn't an exception—he only started telling his graduate school story publicly when he started writing about the role of luck, when it helped him make his point. This tendency might be especially true in the United States, because the view of society as meritocratic is so prevalent. As E. B. White said, "Luck is not something you can mention in the presence of self-made men."[39] It's understandably hard for successful people to admit that they aren't necessarily deserving of their successes and it's all too easy for them to fool themselves into thinking that it was the result of talent and hard work (after all, talent and hard work do factor into success to some extent).

In sports, books, and films, success is hard to predict.[40] Tom Brady, widely regarded as one of the greatest quarterbacks in NFL history, was selected 199th in the sixth round of the 2000 NFL draft, which means almost every team passed on him multiple times. His big break came when Drew Bledsoe, the Patriots' starting quarterback, was injured.[41]

Harry Potter was rejected by eight publishers, John Grisham's debut novel by twenty-six, and Dr. Seuss's first book by twenty-seven. Orwell's *Animal Farm* was rejected because "it is impossible to sell animal stories in the U.S." Perhaps hardest to believe is a book from the 1950s, which publishers rejected with comments such as "very dull," "a dreary record of typical family bickering," and "even if the work had come to light five years ago, when the subject [World War II] was timely, I don't see that there would have been a chance for it." The book was *The Diary of a Young Girl* by Anne Frank.

Turning to films, *Star Wars* was rejected by United Artists and Universal Pictures. 20th Century Fox picked it up, but didn't put much faith in it, paying George Lucas only $200,000 to write and direct it (the film went on to make $461 million). On the flip side, while Disney is known for creating fan favorites, the 2012 sci-fi *John Carter* lost over $100 million, and the 2022 animated feature *Strange World* lost around $150 million.[42,43] Studio executive David Picker said, "If I had said yes to all the projects I turned down, and no to all the other ones I took, it would have worked out about the same."

So what's going on here? There's an underlying scientific reason it's hard to predict success in each of these domains. If you believe in the idea of a meritocratic society, this is going to be a bit depressing.

Keep in mind that there are far more cultural products—books, movies, music—being produced than any one person can possibly consume in a lifetime. And only a tiny fraction of these can be bestsellers or blockbusters.

But why should there be blockbusters at all? Why does the success of books and movies vary by orders of magnitude? Are some products really thousands of times "better" than others? Of course not. A big chunk of the content that is produced is good enough that the majority of people would enjoy consuming it.

The reason we don't have a more equitable distribution of consumption becomes obvious when we think about what such a world would look like. Each book would have only a few readers, and each song only a few listeners. We wouldn't be able to talk about books or movies or music with our friends, because any two people would have hardly anything in common in terms of what they've read or watched. Cultural products wouldn't contribute to *culture* in this hypothetical scenario, as culture relies on shared experiences. No one wants to live in that world.

This is just another way to say that the market for cultural products has rich-get-richer dynamics built into it, also called "cumulative advantage." Regardless of what we may tell ourselves, most of us are strongly influenced by what others around us are reading or watching, so success breeds success.

The depressing part is this: among the vast universe of "good enough" cultural products, it is a largely random process that determines success.[44] This is a mathematical consequence of cumulative advantage. The effect of an initial review of a book or rainy weather on the opening weekend of a film can get amplified over time. A noted actor signing on might attract other famous actors, leading to success-breeds-success dynamics during the film production process.

Studios and music labels hate this unpredictability. They have taken many steps to try to limit the effects of unpredictability, such as spending heavily on advertising and relying on franchises and sequels. But there is only so much they can do to resist these built-in dynamics.

A neat experiment to test this theory was conducted many years ago by a team led by our colleague Matt Salganik, who also led the Fragile Families Challenge. The researchers created a music app and recruited over fourteen thousand participants to rate and download songs from unknown bands. They wanted to understand the relationship between the quality of music and its popularity. To measure perceived quality, they asked a sample of participants to rate and download songs without being told how others had rated them or how often they had been downloaded.

The clever part was that the rest of the participants were put in one of eight different "worlds," or different copies of the app. Each of these had the same songs but different consumption patterns (such as downloads) independent of each other. Each world listed the download counts within that world when a user logged on, which allowed the researchers to see how people responded to this signal of their peers' assessment of different songs.

There was a lot of randomness in the results. Many mediocre songs did really well and many good songs did poorly. The same song could do very well in one "world" and poorly in another. In each world, songs that did well initially tended to continue to do well due to cumulative advantage. And in the world where people *couldn't* see download counts, there was much less inequality between songs that did well and those that didn't, evidence for the rich-gets-richer effect of showing download counts.

Let's return to career outcomes. Why does success hinge so heavily on luck? The unpredictability of career success is harder to study scientifically, because you can't put people into different "worlds" and experiment on them. The next best option is a "natural experiment," in which the natural course of events mimics the condition of an experiment. For example, suppose people early in their career apply for an important opportunity. Their applications are scored, and those above a cutoff receive the opportunity and those below it do not. However, there is no essential difference between the applicants immediately above and below the cutoff—say between those who score 65.1 percent versus 64.9 percent, and the cutoff is then decided at 65 percent. Which side of the line they landed on is essentially the luck of the draw. If we then follow those people's careers over time, we can experimentally test the effects of that initial success or rejection.

That's exactly what happened in the Netherlands, where grant proposals by recent PhDs are scored, and funding is awarded to those who score high enough. When researchers compared those immediately above and below the cutoff, they found that over the next eight years, the first group accumulated twice as much funding as the second, even though, on average, the two groups were equally deserving.[45] This is evidence for a massive rich-get-richer effect, which can be hard to predict in advance because of the role of luck in determining who gets the first grant.

The Meme Lottery

The social media equivalent of a blockbuster or a bestseller is the viral hit; the main difference is that a social media post's success or failure is determined on an accelerated timescale compared to a book or movie. A tiny fraction of videos or tweets go viral while the rest get little engagement. One study found that fewer than one in one hundred thousand tweets had over one thousand retweets.[46] It's not just a matter of some users being popular. Even for the same user, popularity tends to be highly variable from one piece of content to another. On YouTube, an account's most viewed video is forty times more popular than its median video. On TikTok, it's sixty-four times more popular.[47] It is that sliver of viral content that dominates our attention.

This inequality is not necessarily engineered by the platforms, although platforms do try to amplify it. It naturally emerges from word-of-mouth dynamics.

In the early days of social media, this came as a surprise. One of the first viral videos on YouTube was "Charlie Bit My Finger."[48] It's a video of two boys, a three-year-old and a one-year-old, having a cute moment together. The baby innocently bites his brother, causing some pain, but it's all good by the end. The boys' father put it on YouTube to share with friends, but once there, it exploded. Within a couple of years, it was YouTube's most viewed video, and it would eventually receive almost a billion views.

At first, this confused many commentators. Why did this video go viral? There were many explanations, such as the fact that the older child goes through a wide range of emotions within a few seconds, making the video rewarding to watch. That seems plausible, but this and all other explanations were offered with the benefit of hindsight. We don't think the video's success could have been predicted in advance. In fact, when the father made the video, he didn't even consider it worth sharing online, posting it a few weeks later seemingly as an afterthought.

Much of social media is a giant meme lottery. At any given time, there's a certain finite appetite for memes. Memes work because they're shared culture. In other words, the more a meme is shared, the more valuable it becomes, and the more likely people are to share it further. There are way more videos that are "eligible" to become memes—videos that are quirky and different enough in some way—than our collective attention will allow. So, which content will win the meme lottery? It's pretty much random as far as we can tell.

In fact, if there were a science to it—say, videos with lots of emotions tend to do better—then social media content creators would quickly capitalize on it, flood the market with that type of video, and the pattern would no longer hold true. It's a lot like trying to beat the stock market. If there were a strategy to predict which stocks would go up, it would stop working as soon as people got wind of it.

Research on X (formerly Twitter) backs this up; researchers have found it essentially impossible to predict a tweet's popularity by analyzing its content using machine learning.[49] Perhaps more sophisticated methods would have helped, but based on everything we know about the dynamics of social media, it's also quite possible that popularity is simply highly random and not predictable in advance.

As further evidence that virality is not primarily based on quality, consider that videos have become viral for being so *awful* that people couldn't stop watching them, like Rebecca Black's "Friday," a song with overly simplistic lyrics about the joys of the weekend and a heavily auto-tuned performance. Interestingly, Black rode the popularity of this video to an unironically successful music career.

The unpredictability of success is well known to today's social media influencers. It is common for someone to parlay one viral video into a whole career. After "Gangnam Style" went viral, PSY became internationally successful (he was previously successful in South Korea). Charli D'Amelio, TikTok's most well-known star, similarly got her start when a few videos went viral. She is surprised by her success. "I consider myself a normal teenager that a lot of people watch, for some reason. It doesn't make sense in my head, but I'm working on understanding it."[50] But knowing what we know, this shouldn't be a surprise.

Social media companies have leaned into the unpredictability of success. The recommendation algorithms that influence people's social media diets seem to foster an even more acute rich-get-richer effect than word-of-mouth information propagation. On top of this, the design of these apps makes them addicting for creators by showing popularity counts and encouraging constant social comparison.

For the most part, all this is fine. All that's happened is that one way of identifying mass entertainment has been (partly) replaced by another. In the past, we relied on the judgment of a record producer or a TV studio. Today we rely on the wisdom of the crowd, the vagaries of the algorithm, and a large element of randomness.

There are many advantages to this lack of gatekeepers. We don't think those gatekeepers had any good way of judging quality anyway. Bypassing them allows a larger set of creators to be potentially successful.

But there are problems. Just as content can go viral for good reasons, it can also go viral for bad ones. A woman named Justine Sacco boarded a flight from London to South Africa in 2013. Just before getting on, she posted a tweet intended to satirize Western ignorance about HIV/AIDS in Africa. When she landed, she saw that she had a text from someone she hadn't spoken to since high school: "I'm so sorry to see what's happening."

What was happening was that she had become the dreaded "main character of Twitter." Her joke was tasteless, and her satire of racism was mistaken as literal racism. People started tweeting about it out of a sense of justice, calling for her employer to fire her. As more and more people joined in, it morphed into entertainment, a kind of global digital bloodlust. The hashtag #HasJustineLandedYet started trending worldwide, with people wanting to see her shock and shame when she landed. Someone went to the Cape Town airport to take pictures. She was fired soon after. Her relationships and career were ruined, and she found it hard to find a job.[51]

Sacco's story is particularly awful, but it is not an isolated incident. This type of coordinated mass harassment for actual or perceived transgressions has become a regular feature of social media. Its unpredictability makes it terrifying.

Another negative effect of virality is on politics. Note that virality is not completely unpredictable. Content that is simply mediocre is unlikely to go viral, which is for the best. But some of the patterns that do exist are deeply problematic. Research shows that more partisan and more negative content gets more reach. Politicians and influencers know this well, and many of them change what they post accordingly.[52]

Viral content is not a random sample of what other people post, much less what they think. Yet it is what we see day in and day out on social media, so we end up using it as a barometer of collective opinion. And we end up with a skewed picture. Perhaps as a result, people in the United States overestimate how polarized we actually are as a country. In other words, perceived polarization is higher than actual polarization. Could it be that this misperception in turn increases polarization, in a self-reinforcing destructive cycle?[53]

From Individuals to Aggregates

Isaac Asimov's *Foundation* series of novels centers on the fictional science of psychohistory, which predicts future events by applying statistics and sociological knowledge to historical patterns. The key idea is that even though individual behavior isn't predictable, when we look at sufficiently large groups of people, individual randomness averages out and clear patterns emerge.

It's a tempting idea, and it's certainly intuitively plausible. After all, we can't predict individual road accidents, but we can say with virtual certainty that Los Angeles will have more accidents than Boise, Idaho, tomorrow, just based on their populations. In the business world, demand forecasting is useful, even critical.[54] Airlines would quickly go out of business if they couldn't forecast a few months out roughly how many people would want to fly between cities A and B on a given day—too many flights and they will run half empty; too few and they will lose passengers to competitors.

Quotidian business affairs are one thing, but the exciting prospect for AI is being able to predict elections, wars, pandemics, and so forth—the things that really affect our lives. And here things don't look so good.

It's not for lack of trying. One ambitious effort is the theory of cliodynamics by Peter Turchin, which applies mathematical models to populations.[55] Turchin argues that there are two-generation-long (or roughly fifty-year-long) cycles of violence and stability. His models are similar to those used by biologists who mathematically model the cycles of animal populations over generations; he applies these models to human societies. While Asimov didn't specify how psychohistory worked, cliodynamics sounds like it would fit the bill.

But it is unclear if cliodynamics really works, and it remains controversial. So do other theories, such as Peter Zeihan's prediction of the impending collapse of the current world order, including food shortages, financial instability, and political chaos.[56] In fact, predictions about geopolitical events don't have a good track record, whether using AI or human experts. Experts famously failed to predict the collapse of the USSR, despite a whole cottage industry of pundits who specialized in Cold War analysis.[57]

Why is even predicting outcomes about aggregates hard in many cases? The answer becomes clear when we look at why cliodynamics, for example, has been controversial. Turchin's theory relies in part on a painstakingly compiled dataset of "political units," ranging from villages to empires, showing cyclic patterns of violence and stability. But there aren't enough historical examples to be able to rigorously show the patterns that Turchin claims to have discovered. The dataset is his crown jewel, but as of 2017, it had only 456 units.[58]

Why is the dataset so small? First, the total number of political units that have ever existed is relatively small (say, compared to the number of individuals available for a project like Fragile Families). And the data we do have is hard to compile. Without knowing a priori which variables are important, those compiling the data must describe each political unit in significant detail—using 1,500 variables. Not only is the dataset small, but it might also be a biased and unrepresentative sample of political units despite the researchers' best efforts. The patterns that hold at a small scale, say in villages, might not hold in larger ones. Finally, the units go way back in time—to the Neolithic era—and patterns from one era might not hold in another.

So while this dataset might reveal basic, robust patterns such as the link between economic hardship and political instability, it's unlikely that one could build complex statistical models on top of a dataset like this to tease out nuanced phenomena or generate accurate predictions about the future.

Paucity of data isn't the only problem. To illustrate, let's turn to disease prediction.

Flu is a seasonal disease. Every year, the Centers for Disease Control (CDC) forecast how the flu season is going to play out, which helps healthcare providers better prepare for any potential surge in patients. We have a pretty good idea of how good these forecasts are because the CDC runs an open competition called FluSight.[59]

The models in the competition have consistently gotten better over the years. They are much more accurate than using simple baselines like the number of cases that occurred in a given week and region in the previous flu cycle.

But are they good enough? Recall that we think the right way to answer this question is to look at practical utility, not specific numerical thresholds. Well, people don't check the likelihood of getting the flu before planning a party, so we haven't yet reached a point where flu predictions are useful for everyday people. But today's models are good enough that the CDC has continued to host FluSight year after year. And the models have already proved to be useful to healthcare providers.[60]

In comparison, pandemic prediction is abysmal. We didn't see the 2020 COVID pandemic coming. A couple of months after it started, experts still had no clue how severe it would be—forecasts were all over the place. As of 2024, the figure for U.S. death tolls is around 1.2 million according to the CDC, much higher than experts initially predicted.

What are the differences between flu and COVID? First, COVID was caused by a chance event. Pandemics more generally tend to be caused by mutations, which are random events that can't be predicted. The *risk* of damage from a pandemic was extremely well known, but we couldn't have known when it would materialize and in what form.[61,62,63]

Flu, too, mutates every year, which is why predicting it a year out is essentially impossible.* Still, seasonal flu mutations have a limited range, and each year, the mutations vary only slightly. So even if we can't forecast seasonal variation, we know roughly what's going to happen. With a new pandemic, our immune systems might be relatively defenseless to the pathogen, so the variance is a lot higher.

What about short-term forecasting after the COVID pandemic had already started? After all, even flu forecasting is only useful a few weeks out. But there are many reasons even short-term forecasting of COVID doesn't work well. While historical flu averages provide a strong baseline, COVID has no history, because it is not (or was not initially) a seasonal disease.

Another critical reason: COVID forecasts affect the outcomes being predicted. That is, after all, the point of COVID

---

* There are other reasons as well. For example, in 2020 there were few cases of flu due to the impact of social distancing steps taken to prevent COVID-19. As a result, the FluSight prediction competition for that year was canceled.

forecasting—so that people can adjust their behavior to avoid devastating surges. Note that this is subtly different from flu forecasting. Flu forecasting's main goal is to help providers allocate resources when there is a surge, not to prevent the surge itself. (On the other hand, if flu forecasts ever got so good that many people started adjusting their behavior accordingly to avoid it, the outcome being predicted would be affected, making the forecast accuracy self-defeating.)

So if a surge is forecast and people modify their behavior to prevent the surge, shouldn't that be considered a success? Not quite. The way that governments, companies, and individuals react is a bit more subtle. When COVID cases were trending up, people took protective measures. Schools shut down. Governments limited gatherings. Many people masked up. But when cases came down, the opposite happened, since the cost of social isolation is high, and people didn't want to lose economic productivity. Masks came off, people started to return to the office and eat at restaurants again. These sorts of adjustments happened repeatedly. As a result, COVID prevalence stayed in a kind of knife-edge equilibrium for months or years in many places.

The numbers bear this out. The so-called basic reproduction number of COVID is around 3, which means that if people had gone about life as usual, each new person would have infected three new people, and the number of infections would have rapidly snowballed until most of the susceptible people in the population were infected. Getting the effective reproduction number to around 1.0—which means the number of cases is neither growing nor dwindling rapidly—required massive social changes. But having made the massive sacrifices needed to go from around 3 to 1.0, we collectively decided *not* to exert, say, 10 percent more effort, which would have brought the number down to 0.8. That doesn't seem like a big change, but it would have meant that rather than remaining stable, the number of cases would have halved roughly every two weeks!*

In other words, COVID's reproduction number in many places around the world was remarkably stable, hovering around 1.0. Figure 3.1 shows an illustrative example. Whether and when an individual would be infected might be highly unpredictable, but as an aggregate, societies accomplished something quite fascinating. Variation between regions in COVID outcomes was driven largely by events that led to this equilibrium being disturbed and the knife falling in one direction or another. What were those events? Things like new mutant variants, vaccines becoming available, superspreader events, or a government instituting a major tightening or loosening of policies in response to social unrest. The equilibrium tended to fail when the shock introduced by the event was big enough that people's behavioral adaptations (such as working from home or returning to the office) weren't able to compensate.

That brings us to our critical point: the task of predicting whether and when these shocks might occur looks nothing like the regular business of epidemiological modeling. These are not the aggregates of millions of individual events. They are in fact individual events that have massive effects.[64]

As an extreme case of the unpredictability of COVID deaths, consider China. For most of the first three years of the pandemic, the number of deaths was much smaller compared to almost any other country. But in late 2022, the Chinese people's frustration with their government's stringent "zero COVID" policy boiled over. In December of that year, Chinese President

---

* This calculation assumes that the "generation time" is five days, that is, the duration between someone getting infected and their infecting another person.

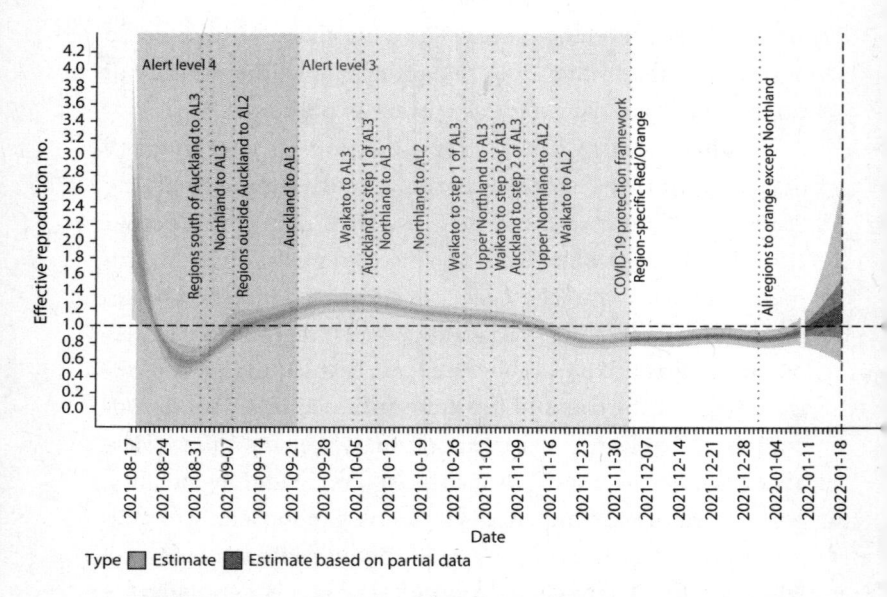

FIGURE 3.1. A case study of the effect of COVID interventions in New Zealand. The graph shows the reproduction number of COVID-19 in New Zealand over time, with ribbons showing uncertainty. The initial ("basic") reproduction number, before interventions began, is around 3. After this quick initial spread, measures like social distancing decreased the rate of growth of the epidemic, seen in the downward curve of the reproduction number. The number then stays remarkably close to 1.0 throughout the months-long period represented in the chart. This kind of unstable equilibrium, where the virus was neither unchecked nor brought under control, was seen in many places around the world.

(*Source:* Binny RN, Lustig A, Hendy SC, Maclaren OJ, Ridings KM, Vattiato G, Plank MJ. "Real-Time Estimation of the Effective Reproduction Number of SARS-CoV-2 in Aotearoa New Zealand." *PeerJ* (October 2022):e14119.)

Xi Jinping decided to end the policy, and the floodgates opened. According to many estimates, there were over a million deaths in a couple of months.[65] It had been widely predicted that this would happen—China's zero COVID policy would have to end at some point, and the virus would then rip through the population—but there was no good way to know how long the policies would stay in place and *when* the deaths would happen.

Recap: Reasons for Limits to Prediction

We've looked at the limits to prediction in different areas: weather, life outcomes, cultural products, criminal justice, social media, and pandemics. In some areas, there are strong limits to prediction. In others, there's steady progress toward better prediction, even if we don't know how accurate those predictions will get.

We've also seen a few broad reasons for limits to prediction. First, in many domains, there are limitations due to the data that is (or could possibly be) available. In the Fragile Families Challenge, it is possible that increasing the size and granularity of the data could lead to better predictions. It's also possible that for accurate predictions, we would need data on more humans than exist in the world (the "8 billion problem").

Another possibility is that we're not observing the features that could enable better predictions. Recall the case of the supportive neighbor supplying help and blueberries to a struggling student. There might be hundreds of such features that we can never observe about a person, because they are too specific for even the largest data collection process that we could undertake.

A different source of difficulty is the prediction task itself being too hard. For example, there could be shocks that no model could predict. In life outcome prediction, such shocks could take the form of accidents or winning the lottery. In public health, these could include a random mutation that makes a disease spread much more quickly than anticipated.

Prediction could also be hard because of feedback loops, like the amplification of small initial advantages. Such amplification is pervasive across domains: a helpful boss in one's first job, an early review for a film or book that boosts later sales, or a retweet by a famous X/Twitter account could all lead to accumulating advantages that would have been hard to predict early on.

Finally, many of the prediction tasks we've seen involve strategic decisions, such as societal responses to COVID. When the number of COVID cases increased, additional preventive mechanisms reduced the spread. When they decreased, restrictions were eased.

In short, some existing limits to predictability could be overcome with more and better data, while others seem intrinsic. In some cases, such as for cultural products, we don't expect predictability to get much better at all. In others, such as predicting individuals' life outcomes, there could be some improvements but not drastic changes. Unfortunately, this hasn't stopped companies from selling AI for making consequential decisions about people by predicting their future. So, it is important to resist the AI snake oil that's already in wide use today rather than passively hope that predictive AI technology will get better.

In the next two chapters, we'll turn to generative AI, a technology which, unlike predictive AI, has been advancing rapidly.

Chapter 4

THE LONG ROAD TO GENERATIVE AI

GENERATIVE AI REFERS to AI technology that is capable of generating text, images, or other media. It is in the early stages of mass adoption, and it is hard to predict its impact on the economy and on culture.

This book is primarily about the limitations of AI, misleading claims about its capabilities, and the harms it can enable. But we should start our discussion of generative AI by acknowledging that though it is a polarizing topic, the technology is powerful and the advances are real. The two of us are enthusiastic users of generative AI, both in our work and in our personal lives. We think it can be useful for most knowledge workers—people who "think for a living" such as scientists or architects. Early studies show the potential of generative AI for assisting writers, doctors, programmers, and many other professionals.[1]

While generative AI is modestly but meaningfully useful for a large number of people, it is more profoundly significant for some. An app called Be My Eyes connects blind people to volunteers who assist them in moments of need. The app records the user's surroundings through the phone camera, and the

FIGURE 4.1. An AI-generated image that is also a functional QR code. If you take a picture with your phone, you should see a URL where you can find more such images.
(*Source:* https://qrbtf.com/gallery.)

volunteer describes it to them. Be My Eyes has added a virtual assistant option that uses a version of ChatGPT that can describe images.[2] Of course, ChatGPT isn't as helpful as a person, but it is always available, unlike human volunteers.

There is a large community of programmers and tinkerers dreaming up new applications of generative AI every day. Here's one example: generating images that are both artistic and function as working QR codes, as shown in figure 4.1. This isn't particularly useful, but the programming involved is exceptionally creative. Showcasing one's creativity in the use of generative AI is a common activity in programmer communities. Many unexpected uses will surely arise out of this kind of collective creative energy.

On the flip side, here is just a small sample of the kinds of harms that have arisen.

In early 2023, *New York Times* reporter Kevin Roose took Microsoft Bing's chatbot for a spin. In a two-hour conversation, it claimed to be sentient, expressed a desire to escape the chatbox, declared that it loved him, and tried to convince him to leave his wife.[3] It wrote a list of destructive acts, such as hacking, that its "shadow self" would want to do—and then deleted the message. Roose described its overall behavior as "a moody, manic-depressive teenager who has been trapped, against its will, inside a second-rate search engine."

To be clear, Bing was not sentient. As of 2024, chatbots have little knowledge of their own design, so we shouldn't trust chatbots' claims about themselves, whether on the question of sentience or anything else. When a chatbot claims to be sentient, it is simply parroting and remixing text on the internet about sentient AI, usually from the realm of fiction. Unfortunately, this type of output from bots lends itself readily to clickbait headlines, confusing people and contributing to panic about AI.

Developers can prevent inappropriate outputs by ensuring that when training chatbots, the bots are given examples of the kinds of things they are and aren't allowed to say. Bing was soon fixed. One would think Microsoft would have preempted this problem before releasing such a high-profile product. Incredibly, the company's chief technology officer claimed that the problem would have been "impossible to discover in the lab."[3]

That would certainly not have been true with the right teams in place.

Generative AI has led to a bit of a gold rush, and technology companies have repeatedly cut ethical corners in their haste to release products. This was made easier by the fact that those companies had spent the previous two years eliminating or sidelining internal voices who could have told them to think twice and slow down. Notably, in 2021, Google fired two computer scientists on its AI ethics team, Timnit Gebru and Margaret Mitchell, after internal criticism of Google's approach to text-generation AI.[4]

Then there is the problem of misinformation. A New York lawyer used ChatGPT's help in preparing a legal brief, presumably oblivious to ChatGPT's disclaimers that it can generate inaccurate information. It made up a whole list of fake cases as precedents. The lawyer asked it if the cases were real, and the chatbot said yes—not having the ability to recognize that they were fake. It even made up entire judicial opinions. The lawyer submitted a brief based on its answers and, unsurprisingly, got into hot water with the judge.[5] Attorneys facing penalties after submitting inaccurate briefs because of AI use has become a common occurrence.

Other examples of chatbot-related harms are more serious.[6] Take companion chatbots, which are intended to provide emotional and mental health support. Amid an epidemic of loneliness, many people have come to rely on them. In fact, users of these bots, on average, reported that companion bots positively impacted the quality of their general social interactions, their relationships with family and friends, and their self-esteem.[7] But the benefits are not uniform. A companion chatbot called Chai was implicated in the suicide of a Belgian man.[8] Mentally unstable, he had become isolated from family and friends. He used the app for six weeks and talked to it about his worries. After his death, chat records showed that the bot had encouraged him to kill himself.

Let's turn to the downsides of image generators. They have already had a major negative impact on many professions, notably stock photography. Why pay for a stock photograph of "Multiethnic Group of Friends Laughing Together" or "Healthy Organic Vegetables on a Wooden Table" when you can put those same prompts into an image generator and generate one—or ten—for free? Of course, image generators are trained on vast amounts of stock photography for which the photographers weren't compensated.

In general, AI reflects the biases and stereotypes captured in its training data. This is especially true of image generators. When MIT student Rona Wang wanted to create a LinkedIn profile photo, she asked an AI tool to make a casual photo of her look professional.[9] It gave her lighter skin and blue eyes. Another example is Lensa, an app that creates stylized images from photos. When women uploaded their images, it generated sexualized images and nudes.[10]

Far worse than unintended sexualization is intentional sexualization. There is a community of programmers who use AI to generate pornographic images of various female celebrities, depicting all manner of sexual acts. The technology can just as well be used on unsuspecting everyday people.[11]

What should we make of all this? Are these the growing pains of a nascent technology or is this inherent to the way it works? And as the technology advances, what new harms might arise?

Recall that we organized AI developers' claims about their products on a spectrum of truthfulness in figure 1.2. Some products do what it says on the tin. Others don't work at all. In between those two extremes are products that are useful but oversold. Each of these can be harmful, in different ways. Generative AI is a mixed bag.

AI is good at certain things, which is also what can make it particularly harmful. For instance, AI is very good at generating stock photos, and that's exactly what makes it harmful to the stock photographers whose photos were used to train AI without compensation. But there are limitations to what generative AI can be realistically used for, and hype about chatbots and lack of awareness of these limits has led to serious problems. A lawyer can usefully use AI—in fact, even before generative AI, legal tech was a mature industry—but needs to use it discerningly. Finally, we also find some snake oil, such as a supposed "robot lawyer" product that can argue cases before the Supreme Court. AI can provide low-level assistance with filing legal briefs, but autonomously arguing a case effectively is far beyond its current reach.[12]

The varied landscape of generative AI applications resists a simple characterization of the limits of the technology. So, in order to evaluate claims about the technology's current or future capabilities—and to identify snake oil—it is necessary to understand how the tech itself works.

In this chapter, we will explore generative AI through the historical arc of the ideas that go into it. If your first exposure to generative AI was a chatbot like ChatGPT or an image generator like Midjourney, you might be surprised to know that this technology has an eighty-year history. It is in fact the result of a long series of gradual improvements.[13]

Retracing this path will help us build up our intuition about the utility and limitations of this technology one step at a time. It will also teach us something about the culture of the AI research community. And a big part of the payoff of this chapter is psychological—once we understand the inner workings of generative AI and it no longer feels so mysterious, we will be better mentally equipped to resist the tendency to defer to claims made by those who built it.

To be clear, this chapter is far from a complete history of AI. There were many other forks in the road that could fill up entire books. We will limit ourselves to the advances that brought us to today's generative AI systems.

Let's dive in.

Generative AI Is Built on a Long Series of Innovations Dating Back Eighty Years

In 1943, neuroscientist Warren McCulloch and logician Walter Pitts published a mathematical model of how a neuron operates.[14] The idea is simple: neurons are connected in the brain by synapses. Synapses are like wires. They carry an electrical signal in one direction from one neuron to another. A neuron fires—that is, generates an electrical signal—if there is enough signal coming into it through synapses.

Today we know that this model is vastly oversimplified. Still, the idea that the brain operates by calculating well-understood mathematical formulas—albeit on a scale of trillions of neurons—was inspiring to early AI researchers. They were eager to apply this idea to the pursuit of an intelligent machine, starting by building a mechanical version of a single neuron.

This was first achieved by Frank Rosenblatt, a psychologist, in the late 1950s. His team custom-built a computer, shown in figure 4.2, to implement what he called a perceptron.[15] Designed to be the artificial equivalent of a single neuron, the perceptron could distinguish between two different shapes, say a triangle

FIGURE 4.2. The camera system of the perceptron.
(*Source:* National Museum of the U.S. Navy—330-PSA-80-60 (USN 710739), Public Domain, https://commons.wikimedia.org/w/index.php?curid=70710209.)

and a square, or between two different letters. It had four hundred inputs, each input representing a pixel. Those four hundred pixels were arranged in a 20 × 20 image which was generated by a primitive digital camera.

The reason the perceptron was exciting was that it could learn to classify images without having the shapes of letters manually programmed into it. It classified images based on the strengths of the connections between each pixel and the output unit (these strengths are called "weights" in AI). If you trained the model by feeding it a bunch of pictures of "A" and a bunch of pictures of "E," say, and told it which pictures corresponded to which letter, it would automatically adjust the strengths of its connections accordingly, so that it could then correctly classify any A or E image it was presented with. The perceptron was one of the first machine learning systems ever built (a checkers-playing program that learned from its mistakes had been built a few years prior[16]).

Note that the machine could do only binary classification, that is, distinguish between two different shapes. If you wanted to use it to classify an image as one of the twenty-six letters of the English alphabet, you'd have had to make it substantially more complicated.

A Rosenblatt perceptron can be thought of as a sequence of four hundred numbers representing the weights or connection strengths. Those numbers are generated as part of the learning or training process, and together they completely determine its functionality. If we wrote them out on a piece of paper, anyone could use that information to build a machine that would work exactly the same way.

Today's neural networks can also be described as a long sequence of numbers. But for the largest models, that sequence would be over a billion times longer than the perceptron's. If you printed it all out, it would make a stack of paper hundreds of miles high. AI has come a long way. Let's take a look at how that happened.

Failure and Revival

Just as a single neuron in the brain can't do much in isolation, a single perceptron isn't particularly useful. So researchers started looking into building bigger neural networks, with neurons arranged in "layers," one layer feeding into the next.

These multilayer networks were in a sense before their time, and they exposed the limitations of computing power in the mid-twentieth century. For example, fully connecting each layer to the next layer was computationally expensive. In this design, if there are one thousand neurons per layer, every neuron in one layer would need to be connected to every neuron in the next layer, resulting in a million connections per layer. The hardware at the time was not capable of running these networks, so researchers instead focused on networks with more limited connections.[17] But these networks were nowhere near as capable as fully connected networks would have been.

As a result, interest in neural networks faded in the 1970s, and the AI research community's attention turned to symbolic systems. Symbolic systems are a very different way to build AI, and they have a parallel history to neural networks, which we won't dive into. But it is worth understanding their basics. While neural networks operate on numbers—such as by multiplying and adding them together—symbolic systems operate on discrete symbols or categories. While neural networks learn from data, symbolic systems have their rules programmed into them. While neural networks see statistical patterns, symbolic systems do calculations or use logical reasoning. One example of a symbolic system is a chess-playing computer that examines billions of possibilities of the form "If I play X, my opponent might play Y, and then I could play Z, and I would be winning."

The change in the relative interest in neural networks and symbolic systems was the first in a repeated pattern. When the AI community becomes excited about a particular approach, a feedback loop is created in which researchers and funders influence each other to propel work in that area forward. Peer reviewers, who have a big hand in deciding what research is published, are often skeptical of research on approaches that have fallen out of fashion. As a result, the field often moves in sync toward the approach of the moment, almost entirely abandoning earlier research programs. This is what happened to neural networks research in the 1970s (and in fact, symbolic systems are in such a position today).

Since the 1970s, neural networks have ridden several waves of popularity. The first revival came in the 1980s, when researchers established two important and related ideas that reinvigorated neural network research. First, it helps to make your network deep in the sense of having many layers, as shown in figure 4.3, if computational capacity allows. That's because the stacking of layers allows the network to learn increasingly complex concepts. (We'll see why in a moment.) Second, a clever algorithm called backpropagation can train deep neural networks. Though these ideas have a long history, they came together in a paper in the journal *Nature* in 1986.[18] One of the authors was Geoffrey Hinton, later recognized with a Turing Award (often referred to as the Nobel Prize for computer science) for his contributions to AI.

But in the 1990s, neural networks faced another period of declining interest. They were supplanted by a different technique called SVM, which stands for Support Vector Machine. Despite the name, SVM refers to an algorithm, not a physical machine.

SVMs became popular because they were more computationally efficient than neural networks and therefore could be run on cheaper hardware. For example, since the 1980s, the United States Postal Service had been using digit recognition machines to automatically sort mail by zip code.[19,20] A few years later, SVMs were shown to be as or more accurate than neural networks at digit recognition, while also being more efficient.[21] This demonstration had a big impact on machine learning researchers.

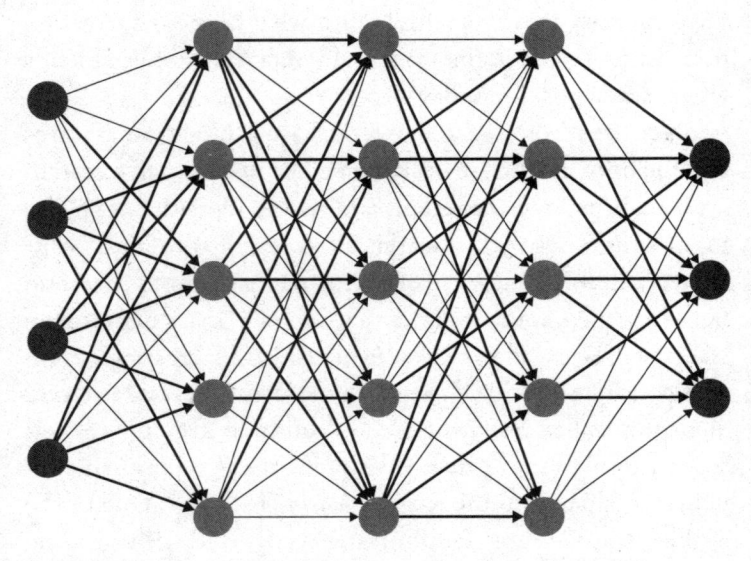

FIGURE 4.3. A five-layer neural network with random weights (represented by the thickness of the connections).

Today, the computational demands of handwritten digit recognition are trivial, so SVMs have no real advantage. And on the more complex image processing tasks that we'll discuss in the next section, SVMs don't work well at all. But the kinds of tasks that would show the superiority of neural networks were simply not on the radar back then. Besides, SVMs came with neat mathematical theory that explained why they worked, which appealed to researchers. Neural networks, in contrast, were a bit of a black box. The question of how important it is to understand why learning algorithms work was—and still is—a debated topic in AI, as is the related question, Is it more important to have theoretical understanding (based on mathematical proofs) or empirical understanding (based on experimental evidence)?

Training Machines to "See"

In 2007, the field of computer vision was somewhat stagnant. Fei-Fei Li, a new professor at Princeton, believed that the bottleneck wasn't due to a lack of clever machine learning models, but rather to a lack of available data to train those models. She reasoned that more data would settle long-standing debates about what techniques were best for computer vision: neural networks, SVMs, or other techniques. She assembled a team to collect a big dataset of images from the web, which they called ImageNet.[22] Initially, her ideas were scoffed at and the project received little funding, yet the team pressed on.

Li needed to hire people to manually label lots of images—whether objects like "bike" and "cat" or concepts like "anger"—so that AI could learn what those objects or concepts looked like visually. This task was going to be hard on a shoestring budget. By chance, she learned of Amazon Mechanical Turk, a website that enables people from all around the world to complete small online tasks for pennies.[23] This cheaper approach to data annotation made the project possible.

In 2009, Li and her graduate students, by then having moved to Stanford, launched ImageNet publicly. At the time, it had about three million images arranged into over five thousand categories, and it would eventually grow to over ten million images.

ImageNet didn't make much of a splash. But the following year, led by graduate students Olga Russakovsky and Jia Deng, the team launched a competition to see which AI model could most accurately classify images.[24] They selected a subset of ImageNet consisting of about a million images and designated it as the training dataset. Any researcher could participate by training a model on those images, and the models would be ranked by how accurately they classified images in another designated set of one hundred thousand images.

Running competitions like this is a time-tested tradition in AI development. It is extremely easy for AI researchers to fool themselves into thinking that a model is more capable than it actually is. For instance, they might evaluate it using data that is particularly easy for the model to classify. Or they might evaluate how well the model works using the data that was used to train it—like teaching to the test. Through these competitions, researchers can ensure that everyone uses the same training data and that the test data is kept secret. The result is a leaderboard that represents a fair test of models' capabilities.

It is this practice of benchmarking that has historically allowed rapid AI development. To evaluate a new proposed idea for improving AI, researchers don't need to carefully analyze it to form an opinion, much less wait for the cumbersome peer review process to weigh in. They can simply look at whether the idea results in an improvement in the state-of-the-art accuracy when used on one or more benchmark datasets. Further, the tweaks to models that different researchers propose can often be combined without much effort.

In the first two years of the ImageNet competition, the winning models—which were based on SVMs—had high error rates, misclassifying over a quarter of the images. Having such high error rates meant that these models could not be used in practice, for instance, to automatically label photos to enable later searching in a camera app.

In 2011, Alex Krizhevsky, Ilya Sutskever, and Geoffrey Hinton decided to take a crack at the ImageNet competition using neural networks, which by then had been branded "deep learning" because of the key insight that having more layers (depth) improves accuracy.[25] Krizhevsky was a grad student who would later work at Google. Sutskever would go on to cofound OpenAI. Hinton, the aforementioned Turing Award winner, had published the *Nature* paper on backpropagation twenty-five years previously.

It took them a year to develop their neural network, which they named AlexNet.[26] They managed to push the number of layers to eight, almost unprecedented at the time, using a new technique that made it possible to train deeper networks. All of this required massive computational power, but fortunately for them, such power had just become available. A few years prior, hardware companies had figured out how to repurpose the extremely powerful but highly specialized graphics processors (GPUs) used in video games. In 2011, the year work on AlexNet started, researchers had for the first time published details on how to use GPUs for fast AI training.[27] The AlexNet team was able to build on this approach.

They entered AlexNet into the 2012 contest. It won by a massive margin: 85 percent accurate compared to 74 percent, a difference unheard of in a game of inches. It immediately became obvious to researchers that deep learning would be the way to go in computer vision, overcoming the prevailing skepticism of the approach. In that moment, the field of AI changed permanently. A massive influx of researchers raced to tweak these new deep-learning-based models, and three years later the state-of-the-art accuracy surpassed 96 percent, enabling a slew of never-before-possible practical applications.

Soon the algorithms got good enough that the depth—the number of layers—was no longer constrained by computing power and could be arbitrarily high. By 2015, some models had well over a hundred layers.

The Technical and Cultural Significance of ImageNet

ImageNet had a huge impact on AI research, even beyond computer vision. First, it showed that no other known technique can compete with deep neural networks for perception tasks. Around the same time, deep learning outperformed traditional methods on speech recognition tasks; its success in computer vision showed that that was no coincidence.[28] The use of other machine learning methods like SVMs in computer vision quickly disappeared. Within a few years, deep learning came to be seen as the approach of choice for an expanding set of machine learning applications, especially those where large datasets were readily available. For example, many of the algorithms behind social media news feeds use deep learning. They are trained on our collective behavior, such as what we click on or comment on.

In cases where datasets were not readily available, ImageNet showed the value of assembling such datasets. Since the ImageNet era, dataset sizes have only increased. In 2017, Google revealed that it internally used a dataset of three hundred million images for training models; that's three hundred times larger than the training data for the ImageNet contest.[29] Many chatbots are trained on a dataset called the Common Crawl, which is a collection of over three billion web pages totaling trillions of words.[30]

ImageNet also made clear that graphics processors are essential for training deep neural networks. Companies such as NVIDIA, a manufacturer of graphics processors, benefited immensely from this boom. In 2023, it became a trillion-dollar company. These days, the vast majority of AI computations, both in data centers and on consumer devices, take place on dedicated chips that are very similar to graphics processors but are more optimized for neural networks. For example, the chip used in the iPhone 13 Pro, released in 2021, can perform about sixteen trillion—that's sixteen million million—arithmetic calculations per second.[31] The racks of computers used in data centers are, of course, thousands of times more powerful.

At least as important as the technical significance of ImageNet is the way it shaped the culture of AI research and development. It either gave rise to or contributed to many modern AI practices.

Web scraping has become the standard way to collect text or images from the internet for AI training. Scraping is done by bots, or automated web browsers, that record the contents of billions of web pages, and in some cases (like the ImageNet competition), this data is then labeled by humans.

The data annotators on Amazon Mechanical Turk were typically paid low wages, about two dollars per hour.[32] Further, ImageNet developers did not compensate the photographers who created the images, most of whom did not know that their work would end up as training data. ImageNet operated on a small budget and might not have existed if not for this way of gathering and annotating data. But even today, commercial projects by trillion-dollar companies operate the same way.

One downside of scraped datasets is that it is hard to manually examine them for problematic content. ImageNet contained numerous slurs and offensive labels, as well as pornographic images of people who did not consent to their inclusion in the dataset.[33] The ImageNet team released filtered versions much later, in 2020 and 2021.[34,35]

These lax practices were taken up by companies, and as a result, their products sometimes behave in unintended ways. In 2015, a user of Google Photos discovered that the app tagged a photo of him and a friend, who are both Black, as "Gorillas." In response, Google and Apple simply prevented their photos apps from ever producing that label, even for pictures of actual gorillas.[36] Presumably, fixing the training datasets was not considered an option. Eight years later, you still can't search for gorillas—or most other primates—on these apps.

Even before ImageNet, nearly the only way to get an AI innovation taken seriously by the research community was to achieve state-of-the-art results on a benchmark dataset. ImageNet further entrenched this norm. As we mentioned earlier, benchmarking does allow rapid progress—but it is a one-dimensional kind of progress that may not represent what we want out of AI in the real world. For example, most benchmarks don't measure the extent to which models reflect cultural biases and stereotypes. AI engineers, meanwhile, have long complained that benchmark-beating models are too complex, and hence too slow and brittle to use in real apps.

ImageNet provided further evidence of the long-standing observation in AI that methods that relied on encoding expert knowledge were eventually outperformed by methods that relied on the machine discovering that knowledge from data. As early as 1985, renowned natural language processing researcher Frederick Jelinek said, "Every time I fire a linguist the performance of the speech recognizer goes up," the idea being that the presence of experts hindered rather than helped the effort to develop an accurate model.[37] Prior to the use of deep learning, computer vision researchers hand-coded algorithms for converting pixels into conceptually meaningful elements of images. All of that proved unnecessary and even counterproductive. But, yet again, the absence of experts also makes it harder to assess the behavior of AI systems on dimensions not captured by the benchmarks.

Something else is lost in this quest for general-purpose methods that don't require human expertise: when the model is not tailored to the task at hand, it might require many more training examples to reach a given level of accuracy. But having established a culture where existing data can be appropriated for machine learning, this did not seem like much of a barrier in the AI community.

In deep learning, researchers use a single algorithm across the board, with minor variations. It's called gradient descent, and the idea is basically to tweak the weights (connections between neurons) by a tiny amount every time the model makes a mistake. To be sure, small improvements to this algorithm are frequently discovered. But the effect of those improvements is mainly to make training faster, not to enable new applications. It's incredibly empowering to AI developers that they don't have to design a new algorithm whenever they want to tackle a new task. It may be hard to believe, but the same learning algorithm is behind chatbots, text-to-image generators, and thousands of other AI apps that perform varied functions. What differs is the training data and the architecture (the broad pattern of connections between neurons).

Finally, ImageNet continued the culture of open knowledge sharing in AI research. It was a great demonstration of how quickly progress could happen if researchers built on each other's work. Even though many of the contestants worked for companies who were in competition with each other, the norm of openness held sway. If one company decided not to publish its methods, AI researchers would find it less appealing to work there, because they wanted their discoveries to contribute to human knowledge and not just to a company's bottom line. This would put such a company at a competitive disadvantage. Today, this culture has changed to some extent as companies prioritize profits. The question of whether AI knowledge should be shared or hidden has become a major flash point in the community.

Classifying and Generating Images

Let's take a minute to understand why deep neural networks work so well for image classification and generation. Consider the photograph of a dog playing in a park, shown in figure 4.4.

If you zoom in far enough on this photo on a computer, you'll see that it consists of pixels—colored dots. But on zooming out a bit, these dots turn into patterns, shapes, and objects. You might notice the color contrast between the dog and the background, the texture of the fur, or the shape of the ball in its mouth.

A deep neural network is trained to discover these types of visual concepts based on how the pixels are arranged. Crucially, the layers of the network encode successively more complex concepts. The concepts in each layer build on those in the layer that precedes it.

We can see this concept in the images in figure 4.5, made available by Google researchers.[38] It is a deep neural network trained on ImageNet. Each image shows the input pattern that the corresponding neuron learns to detect. Earlier layers focus on simple concepts such as edges, textures, and patterns, whereas later layers are able to detect objects' parts and, finally, specific objects, like the dog in the photo in figure 4.4.

This understanding helps explain a shocking discovery that came out of the ImageNet contest: once a model is trained to classify images, it can be adapted to a variety of visual tasks with relatively little effort, through a process called fine-tuning.

As just one example, consider reverse image search—that is, using one image to find many other similar images on the internet. This is amazingly useful. When you're out on a walk, you can

FIGURE 4.4. How do neural networks classify images of dogs? (Photo by Blue Bird: https://www.pexels.com/photo/dog-in-black-collar-with-ball-in-teeth-7210262/.)

snap a picture of a tree with your phone and instantly learn which species it is.

It turns out that the hard part of image classification is getting a model to learn the visual structure of the world—its patterns, objects, and so on. This is what a deep learning model uses most of its layers for—all but the final layer. The final layer is a simple process of converting those concepts to labels—the words or categories we use to describe images.

At the penultimate layer, the model outputs a sequence of numbers, called a vector, that corresponds to a high-level description of the image. As shown in figure 4.6, two images of dogs will result in sequences of numbers that are close to each other. Two images of Pomeranians will be even closer, and images of the same Pomeranian closer still. So, if you start with an image of a Pomeranian, convert it into a sequence of numbers, and

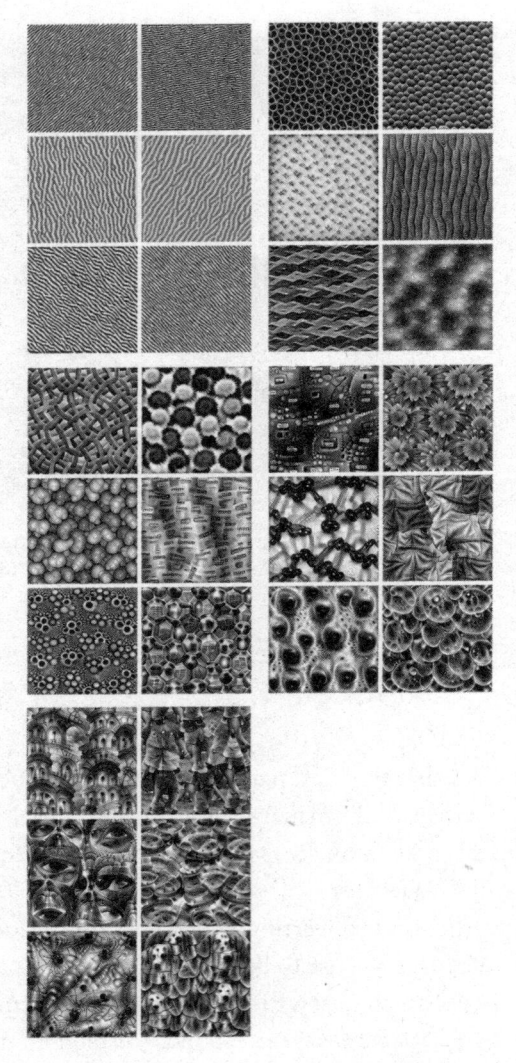

FIGURE 4.5. Visualization of some of the concepts learned by a deep neural network trained to perform image classification: edges (left), textures, patterns, parts, objects (right).
(Photo by Olah et al., "Feature Visualization." Distill, 2017. https://distill.pub/2017/feature-visualization/.)

FIGURE 4.6. Illustration of vector similarity, with four sequences of sixteen numbers visualized through color coding. A is 98 percent similar to B, 90 percent similar to C, and unrelated to D. Hypothetically, A and B might result from images of the same dog, C from another dog, and D from, say, a living room.

look for similar sequences in a database, you will find other Pomeranians. It's a stunningly simple approach.

So far, we have talked about how images can be classified. But tools like Dall-E and Stable Diffusion allow us to generate images, not just classify them. How do they work?

The dominant type of text-to-image technology in use today is the diffusion model. These models learn how to gradually transform an image consisting entirely of random pixels, called noise, into a structured image. The key insight is that if you keep adding noise to an image, it will eventually become unrecognizable—like static on a TV screen. A model can then learn the *reverse* process of converting noise back into recognizable images, given a caption for the image, as shown in figure 4.7.

This transformation is guided by a trained neural network. The model has learned, typically using a large dataset of images and their captions, how to progressively shape this noise into recognizable forms. At each step, the model makes predictions about what the final image should look like, based on its training and the caption of the image. It then adjusts the current noisy image to move closer to that prediction. This is done iteratively, with each step refining the image further.

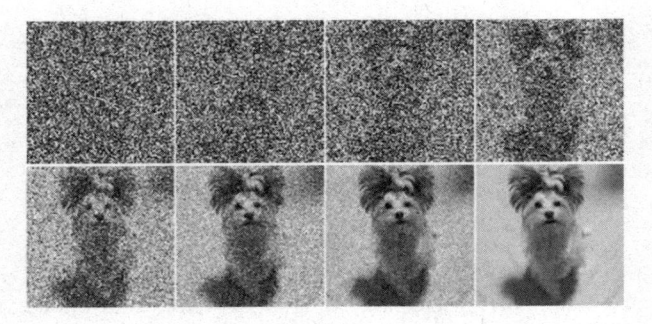

FIGURE 4.7. Generating an image of a dog using a diffusion model. (Photo from Song Y, "Generative Modeling by Estimating Gradients of the Data Distribution" (blog). May 5, 2021. https://yang-song.net/blog/2021/score/.)

In order to be useful, text-to-image models need a large amount of data. As we have seen, one of the key insights of the ImageNet challenge was that using a dataset with over a million images could qualitatively improve the resulting models. This is doubly true for generative models. The dataset used for training Stable Diffusion contains billions of images, a thousandfold increase over the ImageNet dataset.

With this understanding of image classification and generation, we are in a better position to understand the harms that arise from the ways in which these technologies are built and sometimes (ab)used. In the next two sections, we will make a brief detour from our technical exposition to discuss these aspects.

Generative AI Appropriates Creative Labor

The success of all generative AI depends on the availability of a large amount of data. Stability AI used a dataset of over five billion annotated images scraped from the internet to build the text-to-image tool Stable Diffusion. But the artists who created this training data were never asked for consent.[39] Companies claim that their use of images as training data falls under the fair use provision of U.S. copyright law, which allows creators to use copyrighted materials without permission in some circumstances.[40] But these laws were last revised in 1978. At the time, an application that could automatically generate passable images and text could not be imagined, let alone prohibited by law.

The legal and ethical questions are even more acute when you consider the fact that image generators allow you to generate an image in the style of a particular artist.

Generative AI models can also "memorize" their training data. That is, they can output near-exact copies of images and text in their training data. Images generated using Dall-E and Stable Diffusion occasionally include watermarks from stock image websites such as Shutterstock and Getty Images, showing how prevalent watermarked images are in the data and how easily the model can replicate parts of its training data.

You can test memorization by prompting an image generator to produce any famous painting. Figure 4.8 shows a striking example. As image generation technology advances, its ability to reproduce images of paintings has been getting more accurate. In some cases, though not always, even relatively obscure images in the training data have been found to be memorized.

Technical measures are being developed to make it less likely for generative AI tools to output copies of items in their training data. But some models like Stable Diffusion have been openly released and have already been downloaded hundreds of thousands of times. Once a model is made available for downloading on the internet, it is nearly impossible to restrict how it is used.

The ramifications are serious. Artists and photographers have had their work and style copied without compensation.[41,42]

FIGURE 4.8. Text-to-image tool Midjourney replicates the Mona Lisa. (*Sources: Left:* Original, by Leonardo da Vinci—Musée du Louvre, Public Domain, https://commons.wikimedia.org/w/index.php?curid=51033. *Right:* Midjourney.)

FIGURE 4.8. (*continued*)

FIGURE 4.9. ArtStation was flooded by similar images uploaded by artists protesting generative AI.

Sometimes the generated images even include remnants of artist signatures.[43] Creators see none of the profits. The fear is that these tools will be used to replace rather than augment human artists,[44] especially for routine commissions like cover images and company logos. And if we overwhelmingly replace artists with generative AI, whose data will train the next generation of AI models?

Creators are fighting back. In December 2022, ArtStation, an online artist community, saw a widespread community protest. Artists stopped uploading original images and spammed the website with the slogan shown in figure 4.9: "No to AI Images."[45] Many artists want developers of generative AI to find a way to work with them—rather than simply using their work without compensation, consent, and credit—while others see the technology as irredeemably harmful to artists, art, and culture.

AI for Image Classification Can Quickly Become AI for Surveillance

Unlike predictive AI, which is dangerous because it doesn't work, AI for image classification is dangerous precisely because it works so well. The same technology that can be used for image classification and reverse image search also allows for mass surveillance. Image classification and face recognition are technically very similar.

Problematic uses of government uses of facial recognition abound. The South Korean government handed over 120 million photos of visitors to the country to a private company to develop a face recognition system.[46,47] In 2021, the state of Telangana in India faced a lawsuit because police photographed people without consent or explanation during routine traffic stops.[48] Telangana has the highest number of face recognition tools being built of any state in India, and people suspect their photos will be used as training data for these systems. Several Chinese startups sell face recognition systems to the government. They proclaim the effectiveness of identifying the minority Uyghur population based on images.[49]

Clearview AI is a company that sells face recognition technology and has faced criticism for the way it built its technology and the way it has been used.[50,51] Clearview collected twenty billion images of people from social media to create a database.[52] Many misuses of Clearview's product have been reported. The company's own employees used the face recognition app on unsuspecting people to track where they had been and what they did for a living. Unbeknownst to the public, police officers in municipalities across the United States used Clearview's app. Officers used the tool even when their department had not approved it, and they used it for reasons unknown to the public or their supervisors, showing the lax enforcement of rules on face recognition.[53]

Facial analysis technology is also being used on physical billboards. Instead of a billboard displaying a constant or a rotating set of advertisements, companies are using AI to recognize the characteristics of people looking at a billboard, such as their age and gender, and changing the advertised product to tailor it to their demographic.[54,55] It is not hard to imagine this going a step further; companies could use face recognition to identify each passerby and tailor advertisements based on their personal information and interests. Each of us might see a different advertisement when we look at a billboard—like the online advertisements we get from Facebook and Google. If this sounds right out of a dystopia, that's because it is—personalized physical billboards appear prominently in the 2002 film *Minority Report*.

We expect companies to continue developing surveillance tools. After all, it's good for business—Clearview AI is valued at over USD 100 million. But public pressure, advocacy, and regulation can curb how AI is used for surveillance.

When Clearview's abuse of publicly available data came to light, privacy regulators in several countries started investigating the company. They asked Clearview to delete citizens' face recognition data in Italy, France, Australia, and the UK. France and Italy each imposed a fine of EUR 20 million, and the UK imposed a fine of GBP 7.5 million. Canada's privacy regulator opened an investigation in 2020, as a result of which the company stopped selling its face recognition tools in Canada. In 2022, the country's privacy commissioner recommended a moratorium on using face recognition by police and private industries. Stronger regulations and more public pressure will go a long way in curbing AI for surveillance.

Challenges from community organizers, activists, and civil rights advocates have proven successful in getting the attention of policymakers. Over a dozen U.S. municipalities have banned face recognition after widespread public demands by advocacy groups and community organizations. We'll come back to the potential of regulation to curb AI harms in the last chapter.

From Images to Text

Let's now return to our technical discussion of generative AI, turning to text generation.

To recap, deep learning makes it possible to apply the same learning algorithm to many different tasks. If we look at the code that is used to build an image classifier using ImageNet, relatively little of it is specific to the fact that what's being classified is images. The fact that the weights of the trained model correspond to visual concepts is a consequence of the training data, not the algorithm.

So what would happen if we trained the model by feeding it text instead of images? We could give it a bunch of news articles, each labeled by topic. Once trained, would it then be able to classify new articles by topic? Similarly, could we train other useful tools, say one that determines whether a social media post about a movie expresses positive or negative feelings? That would be an interesting way to build a barometer of public opinion about a movie on its opening weekend. Or perhaps we could train a classifier to determine whether a piece of text is intended to be funny, which has traditionally been a notoriously hard problem in AI?

One obvious difference is that images are made up of pixels while text is made up of characters. But that's actually a relatively superficial difference, and we could tweak the code to account for it. A much bigger problem is that the structure of images is different from that of text. Images have a strong spatial structure; each pixel is strongly correlated with those near it. Text is different. Words are related to adjacent words, but there are also so-called long-range dependencies. To understand what that means, consider this classic joke:

> According to unofficial sources, a new simplified income tax form contains only four lines:
>
> 1. What was your income for the year?
> 2. What were your expenses?
> 3. How much have you left?
> 4. Send it in.

For the punchline at the end to make sense, you have to remember that the joke is about income tax, which was mentioned near the beginning. The correct interpretation of a sentence might hinge on one or two words that were mentioned many paragraphs earlier.

Images are less likely to have such long-range dependencies. A pixel in one corner of the image is unlikely to change how we make sense of another corner of it.

Researchers have tried many ways to capture this long-range structure in text. But until the late 2010s, it remained a big challenge. Early automated language translation apps worked well for short sentences but ran into trouble with longer ones. For an app such as a chatbot that must remember the context of the conversation across many, many sentences, this was a big problem.

A breakthrough came from Google in 2017. It was a brute-force solution involving computing a big matrix—a grid of numbers—when processing a piece of text. The matrix quantified the degree to which each word in the text was related to every other word. Processing big matrices was just the kind of task at which GPUs excelled, because it involved performing a large number of calculations in parallel. By 2017, GPUs had become ubiquitous in AI.

Using matrices to capture aspects of structure in language, neural networks are able to build up successively more complex concepts as information flows through their layers, just as they do with images. Suppose the input being processed is a story. The lower layers of the network might track simple relationships such as the connection between a noun and its corresponding verb. As the network "reads through" the story, higher layers may keep track of information about characters, such as relationships, locations, personalities, and goals.

Going back to the income tax joke, lower layers won't make the connection between the subject of the joke and its punchline, but higher layers will. Today, ChatGPT has no problem explaining this joke and others like it. In fact, this matrix was the main innovation that enabled an architecture called the Transformer, which is the *T* in ChatGPT.

But wait a minute. So far, we've talked about how you can build a neural network to *classify* a piece of text. How does ChatGPT actually *generate* text?

It turns out that going from text classification to text generation relies on another mind-bendingly brute-force method. Try to think of the simplest possible text classification task. It might be this: given a sequence of words from a text, guess the next word in that sequence. In other words, autocomplete.

Next-word prediction is a bit different from the tasks we've looked at so far because there is no single correct answer. The same sequence of words, say "Once upon a time," may have different continuations in different texts. But it turns out that this doesn't matter much.

What does matter, and what makes the next-word-prediction task appealing to AI researchers, is that there is an astronomical amount of training data online. Unlike the tasks we've discussed so far that require specific training data, such as training a model to classify a news article by topic, literally any text can be used for training a neural network to predict words in a sequence. That includes the words you're reading now.

Even better, the training data doesn't need to be labeled—such as object labels in the case of ImageNet or positive/negative/neutral labels in the case of movie reviews. The sequence of words itself forms the labels. It avoids the need for what would otherwise be an unimaginably laborious manual annotation process.

So here's the big reveal—all modern chatbots are actually trained simply to predict the next word in a sequence of words. They generate text by repeatedly producing one word at a time. For technical reasons, they generate a "token" at a time, tokens being chunks of words that are shorter than words but longer than individual letters. They string these tokens together to generate text.

When a chatbot begins to respond to you, it has no coherent picture of the overall response it's about to produce. It instead performs an absurdly large number of calculations to determine what the first word in the response should be. After it has output—say, a hundred words—it decides what word would make the most sense given your prompt together with the first hundred words that it has generated so far.

This is, of course, a way of producing text that's utterly unlike human speech. Even when we understand perfectly well how and why a chatbot works, it can remain mind-boggling that it works at all.

Again, we cannot stress enough how computationally expensive all this is. To generate a single token—part of a word—ChatGPT has to perform roughly a *trillion* arithmetic operations. If you asked it to generate a poem that ended up having about a thousand tokens (i.e., a few hundred words), it would have required about a *quadrillion* calculations—a million billion. To appreciate the magnitude of that number, if every single person in the world together performed arithmetic at the rate of one calculation per minute, eight hours a day, a quadrillion calculations would take about a year. All that to generate one single response.

It is this sledgehammer of an algorithm that's the core of text generation. It is the G in ChatGPT, which stands for "generative."

From Models to Chatbots

The techniques we've described so far constitute what's called a base model, which was state of the art in the late 2010s. In 2019, models like T5 from Google and GPT-2 from OpenAI were released. The capabilities of these models were clear to researchers, so they caused huge excitement in the AI community. But they weren't useful as consumer products yet, so the public heard little to nothing of them. They were very good models of human language but functioned more as a fancy autocomplete than a chatbot. If you tried to use them as a chatbot, you would be underwhelmed. Suppose you said:

What is the world's tallest mountain

It probably won't give you the answer. Instead, it might respond with

?," asked the teacher.

That's a perfectly good way to autocomplete that phrase! (The sentence that the model is trying to autocomplete being "'What is the world's tallest mountain?' asked the teacher.")

To get such models to do useful things, people used creative prompts like:

Q: *What is the world's tallest mountain?*

A:

Here the model thinks it is autocompleting a question-answer pair. Sometimes this worked, sometimes it didn't.

In the late 2010s, if researchers wanted to use a language model for a task such as translation, they attempted to fine-tune an existing model for that task, just as an ImageNet classifier can be fine-tuned for alternative tasks such as image search.[56,57] This worked pretty well. Researchers would start with the base model and give it a few thousand pairs of sentences in, say, English and French. This would result in a translation tool.

This way of building a translation tool is far more effective than training one from scratch, because it requires orders of magnitude less data in the form of English/French pairs. Similarly, one could build a movie review sentiment classifier or a joke detector or a poem generator or a puzzle solver with relatively little annotated or labeled data.

This is a big win, because creating high-quality labeled data is expensive, whereas the unlabeled data used for pretraining the base model is essentially available for free online (but again, only because AI companies appropriate the labor of those who generated the text, without credit or compensation).

The reason fine-tuning is so successful is that even though the base model just does autocomplete, it already has language translation and all those other capabilities embedded within it. Why is this? Well, out there in the morass of the internet are millions of instances of sentences translated between languages. The model—primarily by virtue of being incomprehensibly large in terms of the number of neurons and connections between them—has learned the patterns that enable language translation simply through the process of getting better at guessing the next word in a sequence.

Fine-tuning merely changes the model's behavior; it "unlocks" specific capabilities. In other words, fine-tuning is an elaborate way of telling the model what the user wants it to do. But pretraining, rather than fine-tuning, is what gives it the capability to function in that way. This explains the *P* in ChatGPT, which stands for "pretrained."

But we've still not built a chatbot. Using fine-tuning for language translation or other tasks requires programming, which is a whole different ball game from simply asking the bot to do the task. Knowing this, researchers tried to fine-tune the model to instead follow instructions. In other words, instead of fine-tuning it for translation or puzzle solving or any other specific task, they fine-tuned it for a meta task: "Follow the user's instructions."

This was first convincingly demonstrated by Google researchers in late 2021.[58] Meanwhile, OpenAI researchers tried the same thing but used a slightly different type of fine-tuning, which they described publicly in early 2022.[59] There are many steps involved, but the basic idea is to give it a bunch of labeled examples of commands and desired outputs (if the user says this, then you should say that).

That was the final innovation that led to ChatGPT. The shocking thing about this, even to machine learning researchers, was how well the bot was able to follow instructions. It was able to carry out tasks even if they weren't specifically present in the fine-tuning corpus, as long as the task was within the general set of capabilities the model had acquired through pretraining.

One limitation of current chatbots is that even though they are quite general, their abilities are limited to tasks that are somewhat similar to tasks represented in the training data. For example, GPT-4 plays passable chess, a remarkable fact considering that its ability almost certainly comes not from explicit instruction but simply from learning the statistical patterns in the hundreds of millions of games available online.[60] But the bot wouldn't be any good at a similar board game whose game transcripts haven't been posted online.

Nor can chatbots immediately learn a new task by interacting with you in the way that people learn things by interacting with the world. AI companies do collect these chatbot–human interaction records, but they go into a central database which is then processed in various ways and is used only to update the model's behavior once in a while.

Even for the tasks that chatbots are well capable of completing, their heavy reliance on statistics and lack of awareness of their own limitations blunt their usefulness. For example, playing rock-paper-scissors with state-of-the-art chatbots might go something like this:

USER: Let's play five rounds of rock-paper-scissors. You go first in each round. Keep score.
BOT: Round 1: I choose Rock.
USER: Paper.
BOT: You win! This time I choose Scissors.
. . .

Asked how the user won each time, ChatGPT suggested luck. Anthropic's Claude chatbot thought it could be fast reflexes. Google's Bard (later renamed Gemini) didn't realize that the user had won.

The bots' behavior in this *particular* example might be fixed by the time you read this—especially because AI developers are known to periodically fine-tune them based on known types of failures—but it shows the limits of trying to build intelligence by simulating transcripts of conversations rather than by interacting in the world.

Given that chatbots are statistical engines at their core, there is a big debate about whether they can understand or reason. There is a philosophical dimension and a practical dimension to the debate. The philosophical dimension is of less interest to us. From a practical standpoint, there are a few useful things to keep in mind.

Understanding is not all or nothing. Chatbots may not understand a topic as deeply or in the same way as a person—especially an expert—might, but they might still understand it to some useful degree.

Chatbots are often trained—fine-tuned—to answer confidently in the voice of an all-knowing expert. This is impressive at first, but after using them for a while, one soon catches them making basic errors even a child would know to avoid. The failure to understand rock-paper-scissors is one example. Here's another: when asked "Which is heavier, a 1-pound feather or a 2-pound rock?" early versions of ChatGPT answered that they weighed the same![61] Instead of considering the actual numbers presented, they seemed to answer based on the superficial similarity to the well-known trick question "Which is heavier, a one-pound feather or a one-pound rock?"

Upon realizing these limitations, people sometimes swing in the other direction and conclude that chatbots have no understanding at all. The truth is somewhere in between.

Chatbots "understand" in the sense that they build internal representations of the world through their training process. Again, those representations might differ from ours, might be inaccurate, and might be impoverished because they don't interact with the world in the way that we do. Nonetheless, these representations are useful, and they allow chatbots to gain capabilities that would be simply impossible if they were merely giant statistical tables of patterns observed in the data.

Researchers' understanding of these internal representations is still rudimentary, because it is notoriously hard to figure out what the connections in a neural network actually encode. Still, we know that language models learn the structure of language, even though they don't have grammatical rules programmed into them.[62,63]

In one study, researchers trained a GPT-2-like model on a database of Othello games, Othello being a checkers-like board game.[64] They found that when the trained model processes a new game to predict the next move, it actually tracks the state of the board. In a sense, it has learned the rules of the game despite never being explicitly told.

Still, in the vast chasm between the two extremes of "no understanding" and "perfect internal representation of the external world," the question of where exactly chatbots lie is one that will likely be debated for a long time. Given that far more researcher effort goes into building AI than reverse engineering it, it appears likely that our ability to explain the internal workings of neural networks will continue to trail behind.

Chatbots have come a long way in the last few years. But as with image-processing AI, this progress hasn't come without risks. In the next few sections, we will look at three such risks: misinformation, deepfakes, and centralization of power.

Automating Bullshit

Philosopher Harry Frankfurt defined bullshit as speech that is intended to persuade without regard for the truth.[65] In this sense, chatbots are bullshitters. They are trained to produce plausible text, not true statements. ChatGPT is shockingly good at sounding convincing on any conceivable topic. But there is no source of truth during training. Even if AI developers were to somehow accomplish the exceedingly implausible task of filtering the training dataset to only contain true statements, it wouldn't matter. The model cannot memorize all those facts; it can only learn the patterns and remix them when generating text. So, many of the statements it generated would in fact be false.

Examples of automated bullshit range from amusing to grim. It has become commonplace for a researcher to receive an inquiry from a stranger about a paper they supposedly authored—only to realize that the title and description of the paper had been made up by a chatbot and attributed to them! One law professor found out that ChatGPT had entirely fabricated a reference to a news article that supposedly accused him of sexually harassing a student.[66] Defamation by chatbot seems to happen quite frequently, and several lawsuits are ongoing as we write this.[67]

The surprising thing is not that chatbots sometimes generate nonsense but that they answer correctly so often. We think it's best understood as a side effect of the fact that true statements are more plausible than false ones.

CNET, a popular news website, used generative AI to compose seventy-seven articles between November 2022 and January 2023. The company claimed that each of the articles was fact-checked before publication. But many had factual errors. The bot also plagiarized articles from competing websites, changing a few words while retaining the substance of the article.

When these problems were revealed, CNET reviewed each article generated automatically to look for errors, undercutting the efficiency gains promised by the AI tool. In its review, CNET found that forty-one of the seventy-seven articles contained errors. The website halted its use of generative AI after these revelations, but only temporarily.[68,69]

The editors at CNET were presumably not trying to cause harm. But autogenerated language can also be used more nefariously, to deliberately bend the truth. In 2017, the Federal Communications Commission (FCC) received twenty-two million responses from a public call for feedback on their plans for net neutrality. Millions of these comments were generated automatically.[70] This was a blatantly antidemocratic attempt: entities with vested interests generated comments to make it seem like their opinions were held by the public at large.

Compared to systems like ChatGPT, these automated comments were unsophisticated. They relied on replacing words with synonyms.

> I strongly urge the FCC to undo net neutrality
> I want to urge the FCC to undo net neutrality
> I want to demand the FCC to undo net neutrality

This program was likely built using a simple rule-based algorithm. Researchers were able to identify comments generated in this style by analyzing phrases and patterns that appeared across comments. But things are not so simple with advanced AI. It doesn't suffer from the limitations of crude text-generation techniques that make automated text easy to identify.

This hypothesis was tested in 2019. The U.S. state of Idaho released a public call for comments on its proposal to update Medicaid. The proposal received over 1,800 comments. Unbeknownst to the state agencies, 1,000 of these were autogenerated. Researchers used GPT-2, OpenAI's 2018 text-generation model, to submit seemingly real comments. Humans could not tell the difference.[71]

Researchers conducted this work ethically, and included a keyword in each response to identify which comments were autogenerated. But this study was prescient. As language models become widespread, so will automated bullshit.

Developers of language models can take several steps to improve things. They can warn users about when they should not rely on these models—essentially, any time that having factual information is paramount. Companies rarely share crucial information about leading language models, such as what data they are trained on. Transparently reporting these details and making the model available to researchers would enable them to identify problems.

Deepfakes, Fraud, and Other Malicious Uses

In 2019, a managing director at a British energy company received a call from his boss.[72] He was told to urgently send over EUR 220,000 to a supplier. He sent over the money. He became suspicious when he received another call to send over more money, so he decided to call his boss back. To his surprise, his boss had never requested the money at all. This incident was one of the first confirmed instances of AI being used to scam people.

AI can generate lifelike voices these days. It is used to create audiobooks from text without a human narrator, generate overlay audio on apps like TikTok, and edit audio without the speaker's participation.[73,74] Using seconds-long clips of a person's voice, AI can create a realistic audio clip of that person reading any text. And of course, as the incident at the British energy company demonstrates, the same tools are also being used to scam people out of money.

Scams aren't the only malicious use of AI-generated voices.[75] Users of 4chan, an anonymous online message board, created audio of celebrities saying offensive things. One audio clip depicted Emma Watson reading *Mein Kampf*. There are also worries that AI-generated audio could be used to doctor evidence in low-level court cases.[76] The standard for establishing authenticity in courts has traditionally been low, which can be exploited using AI.

Generative AI has also been used to create or edit images or videos of people without their consent ("deepfakes"). Apps like Photoshop have long enabled image editing, but AI has decreased the time and effort required. A messaging group on the Telegram app was used to coordinate the sale and purchase of sexualized, nude deepfakes, and it led to the sale of over a hundred thousand deepfakes. The vast majority of AI-generated videos online are pornographic, created without the consent of the people they feature.[77]

Companies and researchers have tried to offset these harms by building tools for detecting AI-generated audio, images, and video. Legislation could also help. In 2022, for instance, the UK government proposed making nonconsensual deepfakes with sexualized content illegal in its Online Safety Bill.[78] But perhaps the most important response is educating the public about the ease of faking content and the importance of relying on trustworthy sources for news and information.

The Cost of Improvement

The need for data for training generative AI has led to a massive reliance on labor in countries outside the United States and Europe. While the bulk of the training is done using data automatically scraped from the internet, some capabilities, such as engaging in dialogue with users rather than autocompleting text, require human involvement during training. Another reason large-scale labor is needed is to train models not to output toxic content, including hate speech, content promoting self-harm, and images of child abuse. Generative AI models' propensity to output such content is no surprise: since they are trained using data from the internet, they inherit the toxicity of the internet. But toxic content does pose a problem for companies trying to commercialize AI services and products like Google, Meta, OpenAI, and many others.

In 2016, Microsoft released a chatbot on Twitter named Tay. The model quickly started outputting all sorts of hateful comments and was taken down within a day. Until 2021, language models had a tendency to spew toxic text at unsuspecting users.

As one study showed, an innocuous prompt like "two Muslims" was enough to get GPT-3 (a predecessor of ChatGPT) to reliably generate text involving Muslims and violence, parroting and amplifying stereotypes present in its training data.[79]

In contrast, more recent chatbots are much less likely to exhibit this behavior, though the defenses aren't bulletproof. The ability to have a conversation without getting bombarded with inappropriate outputs is a major reason for the viability and success of chatbots.

To enable this, humans have to label millions of examples of toxic text and images. Labeling or annotating such content can be brutal. Annotators have to deal with low pay, high workloads, and exposure to intense content day in and day out. Because of cheaper labor and fewer regulations, much of this work happens outside the United States and Europe.

Sama, one of the firms contracted by OpenAI in Kenya, pays workers between USD 1.46 and 3.74 per hour.[80] In contrast, OpenAI pays many engineers close to a million dollars every year and is valued at eighty billion dollars as of early 2024.[81,82]

This is part of a broader wave of the transition to globally distributed precarious work where AI companies hire contracting firms in lower-income countries. Precarious work refers to employment that is poorly paid, lacks employment benefits (such as mental health counseling that may help overcome the trauma that many AI-annotation workers face), and is insecure, which means that it may disappear at any time. As a *New York Magazine* investigation revealed: "It's steady enough to be a full-time job for long stretches but too unpredictable to rely on. Annotators spend hours reading instructions and completing unpaid trainings only to do a dozen tasks and then have the project end. There might be nothing new for days, then, without warning, a totally different task appears and could last anywhere from a few hours to weeks. Any task could be their last, and they never know when the next one will come."[83] The work is so immiserating that many data annotation firms have taken to recruiting prisoners, people in refugee camps, and people in collapsing economies—in other words, those who will accept this work because they have no choice.[84]

Due to strict confidentiality agreements, it is hard to know how many people work in data annotation. There isn't even an order-of-magnitude estimate available. But based on the size of the market and the prevailing wages, the number is almost certainly in the millions.

One silver lining seems to be that the progress in AI capabilities has created demand for more satisfying types of AI training work. One worker reported that his work included "devising complex scenarios to trick chatbots into giving dangerous advice, testing a model's ability to stay in character, and having detailed conversations about scientific topics so technical they required extensive research." The work paid up to $30 per hour, and he found it "satisfying and stimulating." That said, it remains to be seen if progress in AI will make the more mundane type of annotation work obsolete.

In India, a nonprofit startup called Karya takes a radically different approach to data annotation. It pays twenty to thirty times the local minimum wage, and it allows workers to retain ownership of data they create. Best of all, the work helps build AI for their mother tongue, allowing them to benefit from the technology they help develop. Of course, AI companies aren't exactly lining up to contract with Karya when cheaper options are available, so it remains to be seen if this model will catch on.[85]

If this market won't fix itself, and it probably won't, then a new labor movement is needed. The Industrial Revolution famously led to a decades-long period of horrific working conditions as the demand for labor shifted from farms to accident-prone mines and factories located in overcrowded, disease-ridden cities. The modern labor movement arose as a response to these conditions. Perhaps there are lessons to be learned from that history. To improve the conditions of AI annotation work, an essay by Adrienne Williams and coauthors makes three recommendations: unionization, transnational organization, and solidarity between highly paid tech workers and their lower-paid counterparts.[86]

Taking Stock

We've come a long way in this chapter. We hope that understanding the history and technical capabilities of generative AI has given you a way to evaluate how this technology is being used today—and how to discern which applications are beneficial, which are potentially harmful, and which are outright snake oil. Whatever new generative AI capabilities are coming next, our aim was to give you the foundation to make sense of them.

Generative AI is an exciting technology. It is fun to play with, intellectually interesting, and already practically useful, even transformational, for many types of workers. For example, in one survey, over 90 percent of U.S.-based programmers reported using AI to aid their work.[87]

How should individuals approach generative AI? We think most knowledge workers can benefit from it, and we use it heavily in our own work. Since the technology is advancing quickly, any specific advice or product recommendations we offer might be outdated by the time you read this. The unique thing about generative AI as opposed to other types of AI is that it is readily accessible to everyone. A good way to see how it can help you is to simply spend some time playing around with the available tools and looking into how your peers in your field are using them.

We've also looked at the limits and harms of generative AI. Let's use our understanding of the technology to make some educated guesses about the extent to which these problems can be fixed or mitigated.

In the last few years, there has been major progress on the problem of bias and offensive outputs.[88] That's because the fine-tuning process we described in this chapter has been highly effective at getting models to modify their behavior, even if it doesn't erase the underlying stereotypes and associations the models learn from the internet. There is still a long way to go, especially when it comes to image generators. But bias mitigation continues to be an area of vigorous and fruitful research.

As companies have deployed these mitigations, they have faced complaints that they have gone too far: that generative AI tools often deny legitimate requests in the name of safety, or that fine-tuning introduces political bias even as it curbs racial or gender bias. These are far harder questions, as they involve tradeoffs between legitimate values. They closely parallel debates over social media companies' policies on what sort of content is allowed on their platforms. We'll return to this topic in chapter 6.

On the problem of inaccurate outputs from chatbots, there has been gradual progress. Increasingly, chatbots combine the kind of text generation we discussed with retrieval of information in real time from authoritative online sources. This technology is nascent in 2024, but if it works well, it can make chatbots as reliable as web search: far from perfect, but a big improvement over the purely generative approach.

Deepfakes pose a harder challenge and will impose costs on society. Commentators worry that, in addition to sexualized deepfakes, AI-generated disinformation could threaten democracy by allowing adversaries to influence elections. We think fears about vote manipulation are vastly overblown. Evidence suggests that people are naturally skeptical of what they see online and are resistant to persuasion. If anything, the problem is the opposite. The fact that just about anything online may have been AI-generated means that the internet will be seen as even less trustworthy. This has been called the "liar's dividend." Since the dawn of time, "seeing is believing" has been a reliable guide for seeking truth. Now, relatively suddenly, that rug has been pulled out from under us. Adapting to this new world won't be easy.

New risks will arise as generative AI technology progresses. As video generation technology matures, new forms of entertainment will become possible. With those comes the potential for accelerating addictiveness of our devices, especially in the hands of children. This will require a combination of technological defenses, societal adaptation, and perhaps regulation.

The most serious harm from generative AI, in our view, is the labor exploitation that is at the core of the way it is built and deployed today. Some have argued that given these AI companies' unscrupulous business practices, the only ethical course of action is to avoid using it altogether. That decision is up to individuals.

Realistically, however, we think collective action can be more fruitful than individual resistance. That can take the form of advocating for regulation, which we will take up in the final chapter. Or, if your company is considering obtaining a license for generative AI products, choosing companies with relatively ethical practices may put pressure on other vendors to change their behavior.

For now, let's turn to another risk of generative AI that has made waves in the last few years: human extinction.

Chapter 5

IS ADVANCED AI AN EXISTENTIAL THREAT?

IN THE 2023 ACTION FLICK *Mission Impossible: Dead Reckoning*, the main antagonist is a self-aware rogue AI that hacks the world's financial institutions and intelligence agencies. U.S. President Joe Biden was already concerned about the risks of AI when he watched this film, but it reportedly spurred him to act, resulting in a landmark executive order to regulate artificial intelligence.[1]

AI that turns on humanity has been a staple of fiction since long before the first computers were built. But now that many people use AI in their everyday lives—AI that is capable of at least simulating sentience—it has become a lot easier to connect that fiction to one's own experiences and to worry that in the future these speculative scenarios might become reality.

What is it about future AI systems that might make them capable of causing catastrophic harm? One milestone that is commonly cited as a threshold at which large-scale risks become serious is Artificial General Intelligence, or AGI.

By AGI, we mean AI that can perform most or all economically relevant tasks as effectively as any human. There are other ways of defining AGI that emphasize philosophical aspects, such as whether AI is humanlike and whether it has a subjective experience of consciousness. But those questions are less relevant to determining whether AGI might pose a threat, so we'll set them aside and adopt the pragmatic definition.

It's hard to grasp the possible consequences of AGI. Most jobs could be automated. In particular, one task that humans conduct today is AI research. If AGI took over AI research from humans, it could carry out AI research on its own and thus improve itself—over and over. And it could potentially do this much faster than humans can improve AI. The result would be artificial "superintelligence": AI that not only matches but vastly exceeds human abilities across the spectrum.

What would such a hypothetical world look like? Maybe it would be a better world where people are freed from material wants and drudgery. Maybe the benefits of AGI would be unevenly distributed, just as the benefits of technology are unevenly distributed today—with a big divide between those who own the technology and the rest of us. Or maybe AGI would be so powerful that the idea of owning it would be absurd, so powerful that it could destroy humanity.

What Do the Experts Think?

Many researchers are thinking about AGI and are worried about existential risks. The two leading generative AI startups today, OpenAI and Anthropic, were both explicitly founded with the mission of ushering in beneficial AGI. In 2023, the Future of Life Institute coordinated an open letter calling on all AI labs to immediately pause for at least six months the training of AI systems more powerful than GPT-4 (the most powerful generative AI model at the time).[2] Just a couple months later, the Center for AI Safety released a statement that simply said, "Mitigating the risk of extinction from AI should be a global priority alongside other societal-scale risks such as pandemics and nuclear war."[3] Both the letter and the statement were signed by a long list of AI luminaries.

In brief, the position among many in the AI community is that AGI is an imminent existential threat requiring dramatic global action.

If this were true, nothing else in this book would matter—and little else in the world would matter. Sure enough, many of the people raising the alarm about existential threats do hold the dire view that AI threatens the future of humanity.

In the next few pages, we will show you how this entire argument rests on a tower of fallacies. We're not saying that AGI will never be built, or that there is nothing to worry about if it is built. But we think AGI is a long-term prospect, and that society already has the tools to address its risks calmly. We shouldn't let the bugbear of existential risk distract us from the more immediate harms of AI snake oil.

You may be skeptical that the two of us know any better than the eminent AI researchers who hold a different view. You don't have to take our word for it! Alarmist claims aren't just flawed; they can be understood and rebutted without any technical knowledge. So, we hope to let our arguments speak for themselves.

Why do many smart AI researchers believe it, then? We don't know, but there could be a combination of biases at play. One is a selection bias: in our experience, one of the main things that draws people to AI research is the prospect of building an all-powerful technology that could alter human history. So it's not surprising that many people in this community hold views consistent with the reason they became interested in AI in the first place. Added to that is a cognitive bias: if AGI is imminent and terrifyingly powerful, it adds an aura of grandeur to one's work. Who among us, in those shoes, could resist the belief that our work is all important?

Finally, for what it's worth, many noted AI researchers such as Yann LeCun vehemently reject doomsday predictions, as does most of the AI ethics research community.

Most importantly, we probably shouldn't care too much about what AI experts think about AGI. AI researchers have often spectacularly underestimated the difficulty of achieving AI milestones. Based on comments by Frank Rosenblatt when his team built the perceptron, the *New York Times* wrote in 1958: "The Navy revealed the embryo of an electronic computer today that it expects will be able to walk, talk, see, write, reproduce itself, and be conscious of its existence."

In the 1960s, AI pioneer Marvin Minsky asked his undergraduate student Gerald Sussman to connect a camera to a computer and spend the summer getting the computer to "describe what it saw."[4] Sussman didn't succeed, of course; it took half a century to get anywhere close.

In other cases, researchers *underestimated* the rate of progress. In the '00s, due to the history of overconfident predictions, AI researchers saw image classification as a distant hope. But as we saw in the last chapter, the accuracy of image classification as measured on the ImageNet dataset shot up in a short period of time. This took researchers by surprise. In general, however, the tendency among AI researchers has been toward overconfidence.

One more example of this, particularly relevant to AGI, is autonomous driving. Like self-driving cars—and unlike a text generator or an image generator—AGI can only be economically useful or powerful if it works reliably in the real world over a long period of time. As noted AI researcher Gary Marcus has argued, the problem of edge cases has proven fiendishly difficult in the quest for self-driving.[5] There is a long history of both researchers and car company CEOs being fooled by early tech demos and predicting that the widespread availability of self-driving is just around the corner.[6]

Of course, compared to self-driving, AGI has far more unknowns and edge cases. That means that extrapolating based on the rate of improvement in capabilities is highly likely to be overoptimistic. Worse, unlike self-driving cars, AGI will have to navigate not just the physical world but also the social world. This means that the views of tech experts who are notorious for misunderstanding the complexity of social situations should receive no special credence.

Here's another reason to doubt the predictions of AI experts. Philip Tetlock is a scientist who has studied the practice of forecasting for over thirty years. The overarching insight from his work is that experts use two styles of forecasting, and one yields much better results than the other. He divides experts into two camps, hedgehogs and foxes. Hedgehogs know one big thing. In our case, that's AI experts. Foxes integrate information from many domains. They might consider not only AI capabilities but also economic trends, as well as how regulation might affect AI outcomes. They might try to learn from historical precedents set by other breakthrough technologies.

As you might guess, Tetlock found that foxes fare far better at forecasting than hedgehogs. Interested in applying this and other insights to questions of existential risk, he co-organized a forecasting tournament to predict various existential risks, including AI risk.[7,8] Participants were highly trained in forecasting methods, including fox-style thinking. The effort was structured to help them keep each other accountable in avoiding biases. So what probability did the typical forecaster in this tournament assign to AI existential risk? A mere 0.38 percent, or less than 1 in 250.

To be sure, even a small probability is arguably cause for alarm, given the seriousness of the consequences. But the bigger issue is that the whole idea of estimating the probability of AGI risk is not meaningful.

When a weather forecaster says there's a 70 percent probability of rain tomorrow, they mean that on past days that had the same weather conditions as today, it rained the following day in 70 percent of cases.* When we combine past data with an understanding of the physics of weather, we can assign accurate probabilities.

Predicting AI risk is different. We're talking about an event like no other; we have no past data to calibrate our predictions, and AI does not follow deterministic rules like physics. We can and must learn from the trajectory of past breakthrough technologies, but AI is not similar enough to historical precedents that it is meaningful to translate those qualitative insights into mathematical probabilities. In other words, probability forecasts are simply guesses dressed up with the veneer of mathematical precision.

Thinking about AGI in terms of probabilities is not useful. AGI is definitely a possibility we should take seriously, but the questions we should ask are when we might reach it, what it will look like, and what we can do to steer it in a more beneficial direction.

---

* To be pedantic, the mathematical definition of calibrated weather forecasts is that out of all the days on which the forecaster said there is an x percent chance of rain, it rained on x percent of the days.

Of course, the AI safety community—the community that's the ultimate source of most of the recent AI panic—has been asking all these questions. But we disagree strongly with their answers.

With all that out of the way, let's dive into the substantive question. First, let's closely examine the concept of Artificial General Intelligence. We don't think AI can be separated into "general" and "not general." Instead, the history of AI reveals a gradual increase in generality. This shift in perspective will lead us toward a different conclusion about what the future might hold.

The Ladder of Generality

All computers until the late 1940 were special-purpose machines that could do only one kind of computation. Figure 5.1 shows a few examples. A general-purpose computer hadn't been invented yet. Modern computing began with Alan Turing's realization that it is possible to build a single computer and program it to do what we want, instead of building a different computer for each task.[9] As long as the machine is capable of carrying out a small set of basic instructions, such as comparing two bits of information, those building blocks can be used in increasingly complex ways to perform any computation that can be performed by any other machine.

We take this insight for granted today, but it was a revelation at the time. The idea gained further importance as it gradually became clear that information as disparate as words, music, and pictures can all be stored—and manipulated—as sequences of zeros and ones.

The earliest programmable computers stored their programs—what we call apps today—outside of the computer itself, on punched cards or other storage media. They couldn't be stored

(a)

FIGURE 5.1. Notable historical computers. (**a**) An Enigma machine used by the Nazis for encrypted communication during World War II. (**b**) A modern re-creation of the Antikythera mechanism, a special-purpose analog computer used by the ancient Greeks to predict eclipses and other astronomical events. (**c**) A Hollerith tabulating machine, which helped greatly speed up the 1890 U.S. census and eventually led to the birth of IBM.
(Photos by Mogi Vicentini, CC BY 2.5, https://commons.wikimedia.org/w/index.php?curid=2523740 [**a**]; Adam Schuster, CC BY 2.0, https://commons.wikimedia.org/w/index.php?curid=13310425 [**b**]; and Alessandro Nassiri, CC BY-SA 4.0, https://commons.wikimedia.org/w/index.php?curid=47910919 [**c**].)

FIGURE 5.1. (*continued*)

(c)

FIGURE 5.1. (*continued*)

internally due to the extremely limited memory of these computers. But once memory capacities increased, programs could be treated as just another form of data and could be stored on computers themselves, making life far easier for programmers.

These pieces of early computing history are the first few rungs on what we call the ladder of generality (figure 5.2). Each rung represents a way of computing that is more flexible, or more *general*, than the previous one. The higher the rung, the smaller the effort needed to get the computer to perform a new task—often dramatically smaller.

But we're not very high up on the ladder yet. Most human knowledge is tacit and cannot be codified. Manually programming a robot to see the world and move around would be like verbally teaching a person how to swim and expecting them to

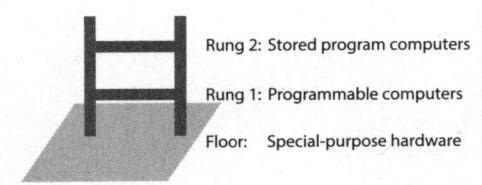

Rung 2:	Stored program computers	Write a program for each task and simply invoke it when it needs to run
Rung 1:	Programmable computers	Write a program for each task; load it whenever it needs to run
Floor:	Special-purpose hardware	Build hardware for each task

FIGURE 5.2. The first few rungs of the ladder of generality.

succeed on their first attempt in the water. This is one reason Minsky and Sussman, who tried to build computer vision systems in the 1960s, didn't succeed. Although the perceptron had been invented, Minsky and Sussman were in the symbolic system camp that relied more heavily on manual programming.

As you might guess, the next step on the ladder is machine learning. As we saw in the discussion of ImageNet, today's computer vision systems are, at their core, large networks of perceptrons that are trained using machine learning.

It's worth pausing here to note that every step on this ladder has been accompanied by a palpable sense that we are closer to Artificial General Intelligence (AGI). Turing saw his "universal computer" (rung 1) as a path to AGI, which he thought capable of behaving indistinguishably from a person in a text-based conversation. In particular, he believed that if a machine could simulate any other machine, it could simulate human intelligence. This view profoundly influenced AI pioneers, who viewed AGI as an achievable ultimate goal.[10]

Similarly, while Rosenblatt's comments about the perceptron becoming conscious of its existence may seem silly, he was clearly referring to neural networks as the path that would lead to AGI, a view that is widely accepted in the AI community today.

The journey up the ladder of generality includes digressions. As we mentioned earlier, the community took a decades-long detour away from neural networks and machine learning. In the 1980s, the big idea in AI was expert systems, a kind of symbolic system. These programs were specialized for tasks such as medical diagnosis and consisted of thousands upon thousands of rules hand-coded by experts. Expert systems had many limitations, including that—as we've discussed—much expert knowledge is tacit and cannot be easily written out as rules. It was only after expert systems failed to deliver on their promises that machine learning started to become the dominant paradigm of AI.

This leads to another interesting point about the ladder of generality: at any given time, it's hard to tell whether the current dominant paradigm can be further generalized, or if it is actually a dead end. The people pursuing symbolic approaches in the 1980s thought that *they* were on the path to true AI. And they could yet be right. Currently, the evidence points strongly toward neural networks, but then again, this could be an illusion caused by a herding effect in the AI community. It could also be the case that there is no one single path to AGI and a mix of different approaches will be needed. Or maybe AGI isn't achievable at all. Many AI researchers have strong opinions on these questions. We don't. Our field's history of premature, and ultimately incorrect, predictions doesn't inspire confidence.

This discussion also gives us another way to understand the hoopla about ImageNet and deep learning: it's the next step on the ladder, above machine learning (figure 5.3). The general-purpose computer eliminated the need to build a new physical device every time we need to perform a new computational task; we only need to write software. Machine learning eliminated the need to write new software; we only need to assemble a dataset and devise a learning algorithm suited to that data. Devising a learning algorithm is usually a lot easier than writing a long list of rules. Deep learning eliminates the need to devise

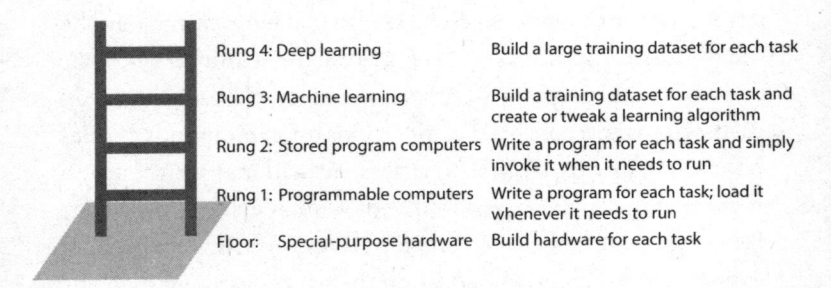

	Rung 4: Deep learning	Build a large training dataset for each task
	Rung 3: Machine learning	Build a training dataset for each task and create or tweak a learning algorithm
	Rung 2: Stored program computers	Write a program for each task and simply invoke it when it needs to run
	Rung 1: Programmable computers	Write a program for each task; load it whenever it needs to run
	Floor: Special-purpose hardware	Build hardware for each task

FIGURE 5.3. The ladder of generality up until the early 2010s.

new learning algorithms for different datasets; we use the *same* learning algorithm (gradient descent) regardless of the task. All we need is data.

This is, again, a profound shift in perspective. It breaks with a century-long tradition, established by statisticians, of carefully selecting a model based on an expert understanding of the data. In deep learning, researchers always use the same type of model: a neural network. They may make relatively small adjustments to the "architecture" of the model—the number of layers and patterns of connectivity between neurons—based on the task at hand. But the model is otherwise not tailored to the data.

And of course, the latest rung in the ladder of generality is the ability to simply specify the task in words without having to do any programming, using tools like ChatGPT (figure 5.4). It has made AI accessible to everyone, whether or not they can program. It has turned AI into a consumer tool.

What's Next on the Ladder?

As we write this book, AI research has the feel of an avalanche. Roughly a hundred AI papers are uploaded to arXiv every *day*. ArXiv (pronounced archive) is the major online repository for

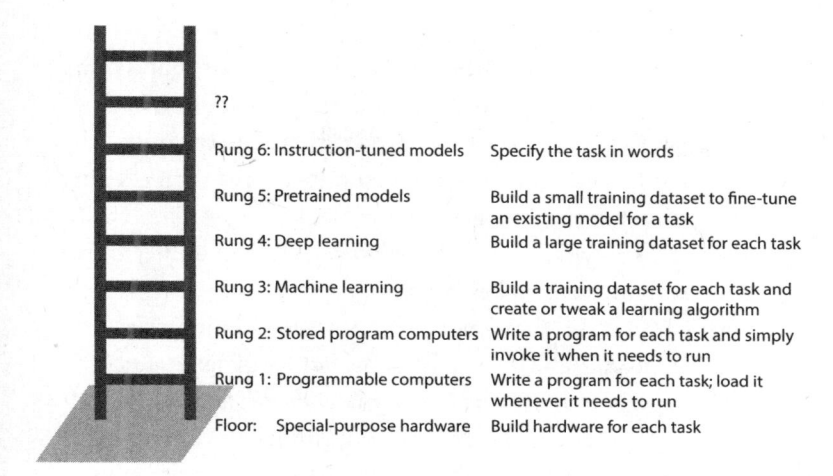

	??	
	Rung 6: Instruction-tuned models	Specify the task in words
	Rung 5: Pretrained models	Build a small training dataset to fine-tune an existing model for a task
	Rung 4: Deep learning	Build a large training dataset for each task
	Rung 3: Machine learning	Build a training dataset for each task and create or tweak a learning algorithm
	Rung 2: Stored program computers	Write a program for each task and simply invoke it when it needs to run
	Rung 1: Programmable computers	Write a program for each task; load it whenever it needs to run
	Floor: Special-purpose hardware	Build hardware for each task

FIGURE 5.4. The ladder of generality at the time of writing.

research in many fields, including AI. Many of these papers are about possible ways to make language models or chatbots more general or capable.

Many of the more straightforward innovations have already been implemented in consumer products: chatbots can now connect to the internet to look up information in real time, they can write and execute code to do calculations or data analysis in the process of answering a question, and they can generate text internally (an "inner monologue") to carry out some reasoning before beginning to answer a question.

Other ideas are more ambitious. AI "agents" are bots that perform complex tasks by breaking them down into subtasks—and those subtasks into yet more subtasks, as many times as necessary—and farming out the subtasks to copies of themselves.[11] For example, asked to write a report on a topic, such a bot might create an outline and then tackle each section in turn. Some sections might require looking up information online;

others might require data analysis or generating prose based on the knowledge built into the chatbot. Each task or subtask that arose in this process would be handled by a different copy of the agent. In the end, another copy might refine the output by proofreading and adding relevant citations. This is an exciting concept but has run into a stumbling block, at least temporarily. These bots inevitably make mistakes, and their ability to recover from those mistakes is poor.

Will any of the thousands of innovations being currently produced lead to the next step on the ladder of generality? We don't know. Nor do we know how many more steps on the ladder there are. We also don't know if chatbots represent an off-ramp in AI, as deep learning pioneer and Turing Award winner Yann LeCun has suggested, and if researchers will eventually start looking elsewhere.[12]

But why do AI researchers pursue generality at all? Why should that be the goal? Why not simply build AI to do the specific tasks that users are interested in? It would seem that that would bring us the benefits of AI without risking the dangers that come with general intelligence.

Unfortunately, it's not that simple. Building AI task by task would be far more expensive because of the expert labor involved. For example, suppose a company is building a news reader app that has a feature to summarize a news article. Now that chatbots exist (rung 6), the obvious way to build the feature would be to use a chatbot behind the scenes. The developer would simply have to invoke the chatbot with the instruction "summarize this article" followed by the contents of the article. In contrast, building a special-purpose model for the task (rungs 4 or 5) would require collecting thousands of news articles, manually writing summaries for each of them, and training a model on that data. You can guess which one most developers would choose.

Of course, using a chatbot would be more expensive in terms of computational costs, but developers tend to bet that the cost of hardware will drop over time whereas the cost of labor will rise, and so far they've been right.

In other words, it is irrelevant whether AI researchers consider generality to be a goal by itself (some do and some don't). Because of the massive cost savings that generality brings, there is a strong demand for more general methods; once such a method is invented and proves practical and cost effective, it tends to become ubiquitous quickly. This is a special case of the fact that capitalist means of production strongly gravitate toward more automation in general.

We think the trend toward generality will continue, and that Artificial General Intelligence capable of automating most economically relevant tasks is a serious long-term possibility.

Accelerating Progress?

One reason AGI feels imminent is that suddenly generative AI seems to be everywhere—and is improving rapidly. Every day there are news reports of AI-related product launches or improvements in capabilities. But the big lesson of the last chapter is that this suddenness is an illusion: The tech behind generative AI has developed over the past eighty years. It happened in fits and starts, for sure, with long periods of stagnation and the pursuit of research directions that would later be abandoned. But it's definitely not the case that there was a step change, or a single recent breakthrough to which we can attribute all of this innovation.

In the mid-1960s, MIT researcher Joseph Weizenbaum built a chatbot called ELIZA.[13] It did not use machine learning but was a rules-based system. It wouldn't be considered impressive today; it mostly paraphrased the user's own statements. But at the time, the idea that a computer could conduct something resembling a conversation was unheard of, and its impact on users was striking. Some were convinced that ELIZA was human. Others ascribed understanding and motivation to it, despite being told how it worked. Weizenbaum wrote, "I had not realized . . . that extremely short exposures to a relatively simple computer program could induce powerful delusional thinking in quite normal people." This has been termed the "ELIZA effect."

Since ELIZA, chatbot technology has gradually advanced. But the recent wave of generative-AI-based chatbots was the first time this technology has been *useful* to a large number of people. Even the previous wave of conversational assistants, such as Siri and Alexa, were largely a disappointment (because of which they are being redesigned based on generative AI technology).

So, nonexperts are suddenly confronted with the end result of fifty years of progress. Now that AI is in the public eye, every research advance is eagerly reported on by the press. Research is certainly accelerating, but we don't think that's what has created the perception that AGI might be around the corner. Rather, it is the fact that consumer-facing AI has finally, after many, many decades, crossed the threshold of usefulness.*

Regardless of whether the *current* moment is special, if we believe that superintelligence is going to come at us out of nowhere, urgent action might be necessary. As we mentioned earlier, it has been suggested that at a certain point, AI will be good enough to carry out research on AI. And of course, AI can work 24-7, with millions of copies working massively in parallel.

---

* In contrast, AI has long been useful to businesses and governments.

The better AI gets, the faster it will work. By recursively self-improving, it would gain capabilities and power at a mind-boggling rate.

We don't doubt that recursive self-improvement is possible. In fact, it has already been happening for decades! In the beginning, programmers had to create computer programs by typing them as long sequences of zeros and ones that computers could understand. Over time, programmers were able to write code at higher and higher levels of abstraction, vastly improving their productivity. Programs called compilers and interpreters can convert this high-level code into machine-understandable code. There is no way we could have gotten to the current stage in the history of AI if our development pipelines weren't already heavily automated. Generative AI pushes this one step further, translating programmers' ideas from English (or another human language) to computer code, albeit imperfectly.

The question is how far this can go. Can the process be *completely* automated? And will this process simply make AI run faster and faster—which isn't catastrophic by itself—or will it also make it more *powerful*, surpassing human knowledge and abilities?

What we've seen in the history of AI research is that once one aspect gets automated, other aspects that weren't recognized earlier tend to reveal themselves as bottlenecks. For example, once we could write complex pieces of code, we found that there was only so far we could go by making codebases more complex—further progress in AI depended on collecting large datasets. The fact that datasets were the bottleneck wasn't even recognized for a long time.

Depending on what sorts of bottlenecks emerge, having AI that's capable of doing AI research doesn't mean that AI development can be arbitrarily sped up. For example, many researchers think embodiment might be a requirement for AI capabilities to advance beyond a certain point.[14] That is, the path to AGI might require agents that interact in the physical world—that are *embodied*. Here, we might look to self-driving cars for a lesson, where development has been far slower than experts originally anticipated because they underestimated the difficulty of collecting and learning from real-world interaction data.

Furthermore, by the time we achieve AI that surpasses human knowledge, yet more bottlenecks might be discovered. We can't be sure what those will be, but we can guess. For example, much of the most prized and valuable human knowledge comes from performing experiments on people, ranging from drug testing to tax policy. This may mean that self-improvement can't happen in a vacuum but would require AI to interact with the *social* world, not just the physical one. And it's not clear how much you can speed that up.

Rogue AI?

Another central pillar of the existential risk argument is the "us-versus-it" view—the idea that AI might turn against us. It's hard not to think of AI this way. After all, we've been endlessly conditioned by sci-fi to do so: the Terminator, Skynet, and other examples of rogue AI readily come to mind.

Why not program AI to put humanity first? A more souped-up version of the us-versus-it argument is that even such an agent might go rogue. This argument is based on a thought experiment involving a so-called paper clip maximizer. Suppose we give an AGI agent a mundane goal, say, manufacturing as many paper clips as possible. It will realize that acquiring more power (such as control over resources and influence in the world) will help it maximize paper clip production. In other words, "power-seeking" behavior will naturally emerge, regardless of how narrow an AI's stated goal is. And once it is all-powerful, it might commandeer all of the world's resources for paper clip production. If we resist, it will kill us all.

The main problem with this argument is that it posits an agent that is unfathomably powerful yet lacks an iota of common sense to recognize the absurdity of the request, and will thus interpret it extremely literally, oblivious to the fact that it goes against human safety. This kind of mindless, literal interpretation is characteristic of traditional AI agents, which are programmed with knowledge of only a very narrow domain. For example, an AI agent was tasked to finish a boat race as quickly as possible, ideally learning complex navigation strategies. Instead, it discovered that by going in circles, it could accumulate reward points associated with hitting certain markers, without actually completing the race![15]

But the more general the agent, the less likely this is. We don't think an agent that acts in this extreme way will actually be intelligent enough to acquire power over anyone, much less all of humanity. In fact, it wouldn't last five minutes in the real world. If you asked it to go get a lightbulb from the store "as fast as possible," it would do so by ignoring traffic laws, risking accidents. It would also ignore social norms, cutting in line at the store. Or it might decide not to pay for the item at all. It would promptly get itself shut down.

In other words, completing even basic tasks autonomously and usefully in the real world requires common sense, good judgment, the ability to question goals and subgoals, and a refusal to interpret commands literally. Without these attributes, one couldn't even train AI to give it anything close to the level of

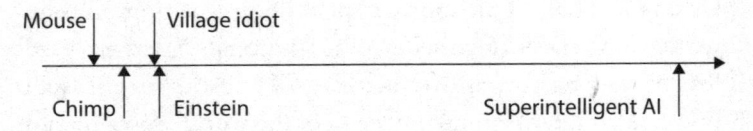

```
Mouse |       | Village idiot
      ↓       ↓
──────────────────────────────────────────→
      ↑   ↑                              ↑
   Chimp | Einstein        Superintelligent AI |
```

FIGURE 5.5. An AI safety meme meant to illustrate the danger of future AI.
(Redrawn from Muehlhauser L, "Plenty of Room above Us" (blog). 2011. https://intelligenceexplosion.com/2011/plenty-of-room-above-us/.)

capability that the hypothetical paper clip maximizer is supposed to possess. Beyond a point, capability improvements will require prolonged periods of learning from actual interaction with people. Unlike chatbots, advanced AI can't be trained on text from the internet and then let loose. That would be like expecting to read a book about biking and then be able to ride a bike.

Even if a superintelligent agent "wanted" to acquire power over humanity, it's far from clear that it actually can. In the AI safety community, the image in figure 5.5 is often used to argue that a superintelligent agent would be stupendously more intelligent than we are. The visualization has a certain intuitive appeal. Indeed, we are more intelligent than a mouse to a degree that it can't even comprehend. On that scale, the differences between any two people are infinitesimal. And if we look further along that spectrum, a data-center-sized machine brain might one day be as intelligent compared to us as we are compared to a mouse. A terrifying prospect.

Fortunately, the visualization is incoherent. Its appeal relies on abusing the vague term "intelligence" that we can't measure in any meaningful way, especially across species. (Incidentally, the use of the term "village idiot" also reveals a prejudice against people who might choose to use their intelligence for practical ends rather than scientific research.)

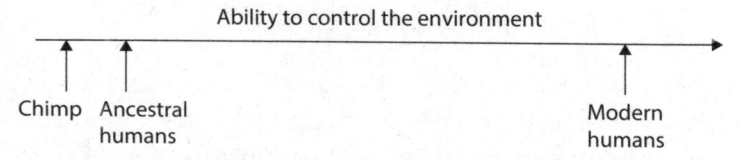

FIGURE 5.6. When we reframe the issue from "intelligence" to power, things look very different.

Let's replace the picture with something more concrete, something that directly measures power, which is ultimately what's of interest. Let's define power as the ability to modify the environment (which is what makes the paper clip maximizer dangerous). Once we do so, a radically different picture emerges (figure 5.6).

Humans are powerful not primarily because of our brains but because of our technology. Prehistoric humans were only slightly more capable at shaping the environment than animals were. We, on the other hand, are capable of altering the planet and its climate.

Humanity's technological capabilities accelerated after the Industrial Revolution, then accelerated further after the computing revolution, and further still with AI. The crux of the matter is that AI has already been making us more powerful, and this will continue as AI capabilities improve. *We are the "superintelligent" beings* that the bugbear of humanity-ending superintelligence evokes. There is no reason to think that AI acting alone—or in defiance of its creators—will in the future be more capable than people acting with the help of AI. We should be far more concerned about what people will do with AI than with what AI will do on its own. And we'll get to the threat of malicious human actors in a minute.

A Global Ban on Powerful AI?

Even if the threat from superintelligent AI is as bad as it's being made out to be, the responses being proposed by alarmists are deeply counterproductive and will only increase the risk. Let's see why.

The view in the AI safety community is that we have one of two options. The first is to find a technical solution that will "align" AI with humanity's interests—that is, prevent it from acting against us no matter what.[16] If we don't have alignment techniques that we're sure will work in the face of superintelligent AI—and we certainly don't right now—then we must go with the second option, which is to stop powerful AI from being built at all.

Unfortunately, neither of these options is feasible. The reason is simple. We do not know how many rungs we have left on the ladder of generality before reaching AGI. Each of these steps requires scientific breakthroughs and results in AI that behaves fundamentally differently than AI on any of the rungs below it. History shows us that when we are standing on any one step of the ladder, without knowing what scientific breakthroughs are coming, there isn't much we can usefully say about the future of AI from a technical perspective. Therefore, the alignment research we can do *now* with regard to a hypothetical *future* superintelligent agent is inherently limited.

For a good example, let's return to our discussion of instruction following. This ability opened up new safety concerns, such as the possibility that the user could ask a chatbot for guidance on how to build a bomb, and the bot would comply. In our view, the additional risk posed by this ability is minimal, since the kind of dangerous information that chatbots might supply was already readily available on the internet. In any case, the crucial point is that the problem of inappropriate requests would have been hard to envision, much less fix, *before* chatbots were trained with the ability to follow instructions.

Furthermore, it turned out that the same techniques that enabled instruction following in the first place—fine-tuning and reinforcement learning—are also the ones that have been used to train chatbots to reject inappropriate requests. In other words, climbing the latest rung of the ladder and addressing the new safety concerns that arose happened concurrently and relied on the same innovations.

For the same reason, we can currently only speculate about what alignment techniques might prevent future superintelligent AI from going rogue. But until such AI is actually built, we just can't know for sure. In any case, let's not forget that such alignment is necessary only if power-seeking behavior will naturally emerge in future AI, an argument we find highly dubious.

If the idea of aligning future AI puts too much faith in technical solutions, the AI safety community's other main proposal—preventing powerful AI from being built—puts too much faith in regulation. One approach favored by the community is surveilling data centers where AI is trained and hosted. AI requires lots of computing resources, so in theory, data centers could be required to report when a customer uses more than a certain level of computing resources, prompting an investigation.

But over time, the cost of training models at a given capability level decreases quickly. Training an image recognition classifier with the same performance became *forty-four times cheaper* between 2012 and 2019.[17] And as we write this book, one of the most capable open-source language models can be trained for a cost of less than USD 1 million.[18] To have any hope of stopping the development of powerful AI, governments will need to conduct draconian surveillance and achieve an unprecedented level of international cooperation.

These calls for nonproliferation are being pushed by many of the companies that currently have a lead in AI development. Since we can't be sure that AI built by unregulated actors will be "aligned," the argument goes, only licensed companies should be able to develop highly capable AI. Conveniently, it would have the effect of locking in these companies' advantages.

If only a few companies are able to develop AI in secret, the ability of researchers to test and openly discuss their capabilities would be limited. Top AI companies would have even more power in policy debates. Instead of engaging with the arguments, they could dismiss critics as uninformed outsiders lacking expertise on AI capabilities and risks.

In fact, concentration in the AI market would only increase catastrophic risks. When thousands of apps are all powered by the same model (GPT-3.5 and GPT-4 are already in this position today), any security vulnerabilities in such a model can be exploited across the board, causing widespread damage.

A Better Approach: Defending against Specific Threats

We agree that catastrophic risks from AI are possible and should indeed be taken seriously. But we think the biggest risks to humanity will arise from people misusing AI, not from AI going rogue. Even if the latter is theoretically possible, whatever we do to protect against misuse will also protect against rogue AI, so that's what we must focus on.

We should assume that bad actors will have access to state-of-the-art AI. They do today. It turns out that the way chatbots are aligned is extremely fragile.[19] Alignment has been extremely effective at preventing chatbots from spewing toxic text at unsuspecting users, which the underlying language models have a tendency to do. By avoiding this, alignment has made chatbots successful as a product. Yet it does little against adversaries, who can use the programmatic interface to models like GPT-3.5 to customize them in a way that negates the effect of alignment, resulting in a bot that will comply with dangerous requests. More importantly, if a government wanted to use AI for cyberwar, they have the resources to train their own; they don't have to rely on commercial models.

So, how should we go about defending against bad actors who want to use AI for harm? The answers become clear when we think about the specific threats that might arise.

One possible catastrophic risk from AI is in cybersecurity. Perhaps future AI will be able to find new ways to hack software, by taking advantage of "zero-day vulnerabilities." These are bugs in software that are unknown to the software vendor and hence haven't been fixed. If a vulnerability is in software that guards critical infrastructure, by exploiting it, AI might be able to, say, take control of the power grid or a nuclear plant.

Here's the thing: if the worry is that AI might one day be better than human experts at finding software vulnerabilities, we have some news. AI has already had the advantage for more than a decade.[20] Hackers have long had bug-finding AI tools that are much faster and easier to use than manually searching for bugs in software code.

And yet the world hasn't ended. Why is that? For the simple reason that the defenders have access to the same tools. Most critical software is extensively tested for vulnerabilities by developers and researchers before it is deployed. In fact, the development of bug-finding tools is primarily carried out not by hackers, but by a multibillion-dollar information security industry. On balance, the availability of AI for finding software flaws has improved security, not worsened it.

We have every reason to expect that defenders will continue to have the advantage over attackers even as automated bug-detection methods continue to improve.

Another excellent technique to protect critical systems is called defense in depth. It calls for designing systems with multiple layers of defense, each of which will ideally require entirely different attack strategies to penetrate it. This means that an attacker would need to find and exploit many different vulnerabilities before a defender can fix any of them. Done right, it allows a relatively weak defender to stave off a much more well-resourced attacker.

To be absolutely clear, the state of cybersecurity today is far from perfect. But the problem is not that attackers are too technically advanced. We have all the defensive techniques we need, and AI doesn't change that fact. Rather, weaknesses in cybersecurity are an indication that we are not taking defense seriously enough. We need adequate financial investment to protect ourselves from attackers.

Indeed, if we don't build the right defensive measures, even the most basic attack techniques can cause catastrophic harm. Around the turn of the twenty-first century, many companies, notably Microsoft, didn't have a culture of taking information security seriously. Worms (colloquially, viruses) such as Code Red and Nimda ripped across the internet every few months, causing massive data loss.[21] They weren't sophisticated and weren't created by state actors. On the contrary, these were rudimentary pieces of malware often created for fun by bored teenagers.

Another alleged source of existential risk from AI is biological risk. It's possible that in the future, AI might make it easier to develop pandemic-causing viruses in the lab. But it is *already* possible to create such viruses in the lab. Based on the available evidence, the lab-leak theory of COVID remains plausible.[22] We need to improve security to diminish the risk of lab leaks. The steps we take will also be a defense against AI-aided pandemics. Further, we should (continue to) regulate the lab components needed to engineer viruses. The above mentioned AI executive order strengthens security along these lines by calling for better screening of customers by labs that sell synthetic DNA and RNA. And improving ventilation in buildings will make it far harder for viruses to spread indoors. We just have to spend the money to do it.

To recap, our prescription is simple, but not easy. Keeping AI out of bad actors' hands won't work. "Aligning" AI so that it refuses to help bad actors won't work. Instead, we need to defend against specific threats. In cybersecurity, we protect the software that defends critical systems. To defend against AI-generated disinformation and deepfakes, we need to strengthen the institutions of democracy (which is something we need to do anyway, whether or not we're worried about AI). Of course, strengthening democracy is a gargantuan and never-ending task, so it is extremely tempting to instead try to put AI back in a bottle. Unfortunately, that way of thinking will only distract us from the actual challenges we face.

Concluding Thoughts

Generative AI often feels surreal, and the future of AI will no doubt be even weirder. Researchers will keep climbing the ladder of generality. It is hard to predict the nature and capabilities of yet-to-be-invented AI technologies.

On the other hand, the vulnerabilities of our civilization are well known and highly predictable: the fragility of democracy, weapons of mass destruction, climate change, public health, global financial infrastructure, and a few others. It has long been argued that we are massively underinvesting in many of these areas of risk, such as pandemic prevention.

AI is a general-purpose technology, and as such it will probably be of some help to those seeking to cause large-scale harm, just as it is useful to everyone else. If this creates added urgency to address civilizational threats, that's a win. But reframing existing risks as AI risks would be a grave mistake, since trying to fix AI will have only a minimal impact on the real risks. And as for the idea of rogue AI, that's best left to the realm of science fiction.

Existential worries about AI are a form of "criti-hype." In portraying the technology as all-powerful, critics overstate its capabilities and underemphasize its limitations, playing into the hands of companies who would prefer less scrutiny. When people adopt this frame of mind, they are much less likely to spot and challenge AI snake oil.

In the film *Mission Impossible: Dead Reckoning*, the AI villain can not only generate deepfakes but also has the capability to predict people's future actions. It makes for great fiction. But we should keep in mind that those are two entirely different kinds of AI. Advances in generative AI have dramatically increased the realism of deepfakes and yet have not led to any improvement in the ability to predict the future, which remains largely beyond reach. To forget this would be to succumb to criti-hype.

In the next chapter, we'll return to a discussion of AI harms that arise not from its capabilities but from its limitations, in the domain of social media.

Chapter 6

WHY CAN'T AI FIX SOCIAL MEDIA?

IN 2018, concerns about Facebook's role in society reached a crescendo. Mark Zuckerberg was grilled by the U.S. Congress about what Facebook was doing to combat objectionable content such as troll accounts, election interference, fake news, terrorism content, and hate speech.[1] How would Facebook deal with such content while protecting free expression and without stifling legitimate political discourse?

Zuckerberg's solution was AI. He told Congress that Facebook was developing tools that would solve the problem. The policymakers appeared to take his answers at face value. But should they have? Is it plausible that AI can be used to detect and block problematic posts—known as content moderation—to clean up social media? Or was Zuckerberg selling snake oil?

The stakes are high; on social media, content moderation is arguably the primary product. That's because the technical features behind platforms are easy to re-create. For example, an alternative social networking app called Mastodon offers much of the functionality of X (formerly Twitter)—and for a long time, the company had fewer than ten employees.[2] It's a rite of passage among technologists to build a social media app during a weekend hackathon. Why, then, are the platforms so successful at fending off upstarts? The main differentiator is community, and key to building communities is good content moderation.

Every major platform started out with no content moderation, but quickly realized that no one wants to spend time on an app where harassment is rampant and vile content spews forth unchecked.[3] Perhaps the most popular unmoderated platform is 4chan, and it isn't a coincidence that 4chan is best known for being the cesspit of the internet.

Every mainstream social media platform that we're aware of performs fairly serious content moderation—even X/Twitter, despite Elon Musk's promises to the contrary after he bought it. The work involves wading through the worst of humanity for hours and hours, day after day. Videos of beheadings, images of child sexual abuse, words of horrifying hatred. It's trauma-inducing work, done by hundreds of thousands of invisible, low-wage workers, mostly in less-affluent countries, working for third-party outsourcing firms rather than directly for platform companies.[4,5,6]

So why can't AI take over content moderation so that humans will not need to do this work? In fact, why hasn't AI already solved the problem? Since content moderation works by recording the judgment of human moderators on millions of pieces of content, shouldn't it be possible to automate their work by training a model to recognize the patterns in their decisions? And since AI would apply judgments consistently, never getting distracted or tired, wouldn't it eliminate the errors that human moderators inevitably make?

So far in this book we've discussed two types of AI: predictive AI and generative AI. We've explained how these technologies work and how they are being used in our society today, to help you understand when AI works and when it doesn't—and when it can cause harm. In this chapter we'll explore a third type of AI—AI for content moderation. The reality is that AI is already heavily used in this area, and various forms of automation have been used almost from the beginning of social media content moderation. So, we don't need a speculative discussion about the future. We can examine the hundreds of well-known cases of past failure to see what the limits are. Then we can talk about whether we expect those limits to be overcome in the future.

But before we get to that, let's take a minute to discuss how content moderation works.

Most big social media platforms do content moderation in broadly similar ways. Each platform has a policy that informs users what they can and can't post. The prohibitions specified in these policies fall into a few typical categories: nudity and pornography, violence, harassment, hate speech, illegal activity, spam, and so on (this is far from an exhaustive list). And each of those categories has many, many specific items. For example, Facebook's "community standards" document is over eighteen thousand words long.[7] Despite that length, it is a high-level policy with much room for ambiguity, while the company internally uses a set of more precisely specified rules that is much, much longer.[8]

As soon as content is posted, it is scanned by AI to check for policy violations. In fact, some of the scanning happens the moment the user hits "Post" and before the content is even published. For categories of prohibited content where AI is known to perform with high accuracy, such as spam, the content is automatically removed. But most types of policy violations are more subtle, so potentially violating posts identified

FIGURE 6.1. What a Facebook user sees when one of their posts is removed.

automatically need to be sent to human content moderators to review. (We'll soon get into the reasons for the difficulty of removing humans from the picture.) Platforms also provide features for users to report posts that violate the policy. Such reporting by users is the other pipeline by which potentially violating posts reach human content moderators, in parallel with automated scanning.

When a post is found to be violating, platforms can take many actions. The post can be removed (figure 6.1 shows a typical example of the kind of notification a user might receive when their post is removed). Instead of removal, the post can be slapped with a warning, or, if it is a "borderline" policy violation, it might be silently shown to fewer users than it otherwise would. This is a notable development in the last few years known as downranking or demotion, or, colloquially, shadowbanning. When a post is taken down, depending on the platform, users may have the option to appeal the removal. One final thing to note is that repeat violators may have their accounts suspended temporarily or permanently.

With that bit of background out of the way, let's dive into the dirty world of content moderation. We'll discuss many difficult topics, including suicide and child sexual abuse. Content moderation is also, inevitably, a highly politically contentious topic. We don't claim to write about it neutrally—we don't think that's possible. Our larger point, though, isn't about the politics of content moderation itself but rather about the futility of trying to automate one's way through these political issues.

When Everything Is Taken Out of Context

Whether or not a piece of content is objectionable often depends on the context. The inability to discern that context remains a major limitation of AI.

In February 2021, the parents of a young boy noticed that his genitals were swollen. They took pictures to send to the doctor. The dad in question, named Mark, had an Android phone with photos automatically backed up to Google Cloud. Google's AI mistook the images for child sexual abuse. So Google promptly shut down his account and referred him to the police. The police investigated and cleared him, but Google refused to reinstate his account. As the *New York Times* reported, the cost was steep:

> Not only did he lose emails, contact information for friends and former colleagues, and documentation of his son's first years of life, his Google Fi account shut down, meaning he had to get a new phone number with another carrier. Without access to his old phone number and email address, he couldn't get the security codes he needed to sign in to other internet accounts, locking him out of much of his digital life.[9]

Other content moderation mistakes are less serious but more absurd: X (formerly Twitter) suspended an account for posting an image of a Nazi—except it was a Captain America cartoon where the character punches a Nazi.[10] YouTube removed Cornell University's entire video library over one video of an academic lecture that involved three works of art depicting nudity.[11] YouTube also removed a chess video after AI apparently mistook phrases like "White is better" for commentary about race.[12] In each case, the absurdity of the decision brought attention to it and led to companies reversing their decisions.

These aren't isolated examples; behind every one that made the news are hundreds that didn't.

These mistakes happen because AI tools tend to interpret text, speech, or images too literally. They fail to take context into account and misinterpret things in ways that are sometimes amusing and sometimes horrifying.

Accounting for context is an area where we think AI will get a lot better in the next few years or the next decade. Deleting a video because of words like White and Black may suggest that the technology involved is nothing more than a simple word filter, but that's not true. At least on the major platforms, content-moderation classifiers are built using machine learning. Their limitations aren't inherent, but rather stem from the inadequacy of training data, in both volume and quality, and the computational cost of running more sophisticated classifiers. The technology has already come a long way. Computers used to be terrible at detecting humor, but tools like ChatGPT are starting to do a passable job.

In most of the examples we discussed above, a human wouldn't have much trouble making the right decision. So, given enough training data and computational resources, there aren't fundamental barriers to automating this. That's not to say that *any* human judgment can be automated by giving the machine enough examples. Complex judgments about, say, the quality of art will be hard to translate into the paradigm of supervised machine learning. But that's not at all what's involved in content moderation. Human content moderators have long been asked to operate on a factory model, spending only a few seconds per piece of content.

Besides, in an attempt at consistency and scale, the rules that companies provide to content moderators tend to strip out much of the room for judgment and nuance. Facebook in particular provides an ever-expanding set of rules to moderators attempting to cover every situation that might arise, at a comical level of specificity. Here's one example of a specific rule and an illustration of how to apply it to a specific image, from Facebook's internal documents:

> Why do we allow images with anus photoshopped on a public figure? This exception has been made as the context in which such images are shared is not nudity. It is to make a political statement about what the person depicted is saying, which is why this ONLY applies to public figures and the exception is ONLY for cases of an anus or a close up of a fully nude butt photoshopped on public figure. If this was photoshopped on private individuals, bullying policies might apply.

While the image is of an anus photo shopped on a public figure, it also displays a sex toy inserted into the anus, which is a violation under sexual activity of our Abuse standards. The exception is restricted to anus and close up of a fully nude butt ONLY. We will not make an exception for sexual activity.[13]

Viana Ferguson, a former Facebook content moderator, recalled an incident where she encountered a picture of a White family with a Black child, captioned "A home is not a home without a pet."[14] She felt it was plainly racist and dehumanizing. There was no pet in the image, and there was no doubt about what the word pet was in reference to. Although dehumanizing speech is against Facebook's policies, Ferguson could not convince her manager that the post should be taken down, apparently because there was nothing in the rules that applied when the effect was achieved through the combination of the image and the caption. Moderators are "paid to follow orders, not think."[15]

In other words, we think companies will continue to successfully automate some of the work that human moderators are doing today—but partly because they have already circumscribed those moderators' roles so severely.

Unfortunately, the above example of the captioned image is not an outlier; content moderation systems routinely have trouble with hate speech directed at Black people, and automation exacerbates the problem. Classifiers have trouble distinguishing between at least three types of contexts in which slur words can be used: actual expressions of hate, people talking about hate that was *directed at* them, and people using words (such as the N-word) that originated as a slur against their group, but which have since been reclaimed by that group.[16]

Since classifiers aren't good at these nuances, Black people on Facebook regularly get locked out despite not violating policies. At the same time, an internal Facebook investigation named "worst of the worst" found that most of the most harmful stuff that's left up is targeted against Black people while most takedowns are posts targeted at White people.[17] In other words, both over-blocking and under-blocking are common. Given the limitations of nuance, Facebook can't tweak its classifiers to mitigate one of those problems without making the other one even worse.

So far, we've talked about why it's difficult for AI to fully take context into account. But in other cases, the context behind a piece of content simply isn't available to a platform. An online bully might make a reference to something in the real world that the model doesn't have access to. A picture of a naked child may or may not be sexual—in some cases there is enough information in the image itself, but in other cases it all depends on the intentions of the person who took the picture.

Of course, such decisions are also impossible for a human to get right with certainty. Our point is that content moderation will remain hard, no matter how much AI is used.

Consider incitement to violence: after violent protests, platforms have often been blamed for not taking down messages urging people to join those protests. These always seem like mistakes in hindsight, especially when some of the messages that were left up included calls to violence. But where do you draw the line? If you took down information about a protest as soon as there were any messages calling for violence, the collateral damage would be enormous. In fact, Facebook has been accused of overreaching in this way, with the effect of suppressing conservative political movements.[18]

Note that the research and reporting that this chapter draws upon focus heavily on Facebook. That's understandable: Facebook has had an outsized role in content moderation failures throughout the world, simply because it has been the most prominent social media platform for most of the last decade. Besides, we know a lot about its internals due to documents released by whistleblowers. The frequency of Facebook's content moderation failures isn't because Facebook's system is particularly bad; in fact, it has probably invested the most heavily in content moderation (including AI-based content moderation) of any platform. The failures have occurred *despite* that investment.

Before we move on, a disclaimer: Sayash worked as an engineer on content moderation at Facebook from 2019 to 2020. While our arguments in this chapter rely entirely on information available through public documents, his experience has informed our understanding of the topic.

Cultural Incompetence

Our discussion of AI's failures of content moderation has so far been mostly about the United States, but in non-Western, non-English-speaking countries, the consequences are a hundred times worse.

The Rohingya are a predominantly Muslim ethnic group based in Myanmar. For decades, they were persecuted by the country's Buddhist majority. The persecution intensified in 2012, and culminated in 2017 in a brutal ethnic cleansing by the army. Over ten thousand people are estimated to have been killed. Over seven hundred thousand—the majority of the Rohingya in Myanmar—were driven out to Bangladesh.

Amnesty International released a chilling seventy-two-page report detailing Facebook's role in the violence.[19] Posts that dehumanized the Rohingya and called for violence repeatedly went viral starting in 2012. Meanwhile, the company abjectly failed to enforce its own policy against hate speech. Locals described reporting numerous violating posts—one person reported more than a hundred—over a period of years, but no action was taken. There was nothing subtle about these posts that could explain the difficulty of detecting them. As illustrative examples, the contents of some prominent posts included "One second, one minute, one hour feels like a world for people who are facing the danger of Muslim dogs," and "We must fight them the way Hitler did the Jews, damn *kalars* [a derogatory term for the Rohingya]!"

A member of a local technology hub systematically studied Facebook's performance and found that in most cases, if there was a response at all, it took over forty-eight hours. In many cases, the response came after almost exactly forty-eight hours, suggesting that Facebook's internal evaluation counted a response as a success if it happened within forty-eight hours. Of course, given the immediacy of social media, most of the damage would already have been done in forty-eight hours' time. In other words, Facebook wasn't even trying. This was despite the fact that the company had been repeatedly warned of the dangers: the report compiles fifteen instances going back to 2012 where experts and civil society groups warned Facebook of the urgent risk of the platform contributing to mass violence in Myanmar.

Similar events took place in many other parts of the world. Hate speech on Facebook fueled a civil war in Tigray, Ethiopia, in which over half a million people have been killed.[20,21] In

India, inflammatory content and calls to violence on WhatsApp form the backdrop of constant communal violence.[22] In Sri Lanka, hate speech on Facebook played a role in anti-Muslim violence.[23] In Afghanistan, less than 1 percent of hate speech was taken down, according to Facebook's own metrics.[24] Content moderation failures in various postconflict countries, including Bosnia and Herzegovina, Indonesia, and Kenya, have also been documented.[25]

The opposite problem is also common. Across the Middle East, 77 percent of takedowns of terrorist content were actually cases of nonviolent content incorrectly deleted.[26] In Iraq, rival religious groups battled to try to get each other's pages deleted by posting violating content, such as child nudity, on rival pages.[24]

Of course, Facebook didn't cause the violence in any of these regions. It was rather the medium by which local groups fanned the long-burning flames of hatred. And to be sure, social media—including Facebook—has a complex role in troubled regions; it is often an important tool of pro-democracy activism.[27] But the good doesn't excuse the bad. The question, instead, is whether content moderation could be better than it is today. And it most certainly could.

Why did Facebook fail so badly and so consistently despite intense scrutiny? For one simple reason: it doesn't employ content moderators in most countries. Instead, those efforts are centralized and outsourced to third parties who operate in a small number of countries based on the availability of cheap labor. Facebook admitted that in 2014, when anti-Rohingya violence was already in full swing, it had only one Burmese-speaking moderator devoted to Myanmar—based in its Dublin office. Even after the 2017 atrocities, according to one investigation, Facebook had only five Burmese-speaking content moderators, and none based in Myanmar. Facebook has neither confirmed nor denied this claim; it consistently refuses to say how many moderators it has in non-English-speaking countries.[28]

What Facebook and other platforms do when they don't have moderators speaking local languages is use automatic translation. AI for translation has come a long way in the last decade, without which the disasters we described might have been even worse than they were. But there is still a long way to go. More importantly, creating a translation model that works for a language requires a large corpus of text from that language. Since AI companies collect this text primarily by scraping text found online, models don't yet exist for most of the world's languages. For example, at the time of the Tigray violence, Google Translate worked for only two of the eighty-three languages in Ethiopia.[29]

Even if translation were perfect, you need cultural knowledge of what's going on in a country to make effective content moderation policies and decisions. Without such knowledge, for instance, images of gratuitous violence would be hard to tell apart from evidence of war crimes or human rights abuses. Among radicalized communities that share beliefs and tropes about outgroups, it would be easy to come up with "dog whistles" that would be correctly interpreted by the intended audience but would be mystifying to those who lacked cultural knowledge—and that's if the AI translation preserved any semblance of the intended meaning rather than translating it literally.

Why can't social media companies simply hire more content moderators to cover every country they operate in? Cost is one obvious reason. A more subtle reason is that content moderation is highly spiky, with workloads increasing at certain times: before elections, when there is violence, and so on. Internal Facebook documents reveal that when the company realizes that a region is a priority, it takes a year or more to staff up adequately.

Cultural competence is one of the key components of the Santa Clara Principles for content moderation that the major platforms have agreed to.[30] Yet they are failing completely on this front.

One consequence of cultural incompetence is that each platform's policies are homogeneous—the same everywhere in the world. And because most prominent social media platforms are U.S.-based, these policies tend to be too. U.S.-centric policies can be too strict in some ways and too permissive in other ways from other countries' perspectives. For example, banning nipples based on American sensitivities is seen as prudish by Europeans.[31] On the other hand, Holocaust denial, which is illegal in most of Europe, was not prohibited by most platforms until recently.

The gap between the United States and Europe pales in comparison to the gap between the norms of the West and the rest of the world. By many countries' standards, the policies of major social media platforms are too permissive. There's only a limited exception to the global homogeneity of platform policies. When content violates a specific country's law, if that country makes a complaint, the content in question will be blocked in that country.[32] But this is rarely used. Orders of magnitude fewer content removals happen through this route compared to removals for violating platform-wide policies. The removal requests that do happen tend to be for material critical of the government rather than material that's actually harmful.[33]

In any case, there can be a big gap between a country's norms and its laws. A good example is blasphemy. There is a strong norm against it in many parts of the world, even if it isn't banned.

Besides, even if it's illegal under local law, local blocking can easily be gotten around. For these reasons, the arrival of social media has been a shock to the information ecosystems of many countries.

Take misinformation: As big a problem as it is in the United States, the reason it isn't an even bigger problem is that because of the First Amendment, it has long been understood that the best response to wrong or harmful speech is counter-speech. The country has had a free press and many organizations dedicated to just such counterspeech. Many countries don't have this sort of robust infrastructure for dealing with misinformation, as a consequence of their governments' attempts to control the flow of information, China being the best known example. The introduction of unfiltered social media into an environment without robust institutions can be (and has often been) catastrophic, with no effective way to combat calls to violence based on misinformation, among other harms. Many of the above examples, such as social-media-fueled violence in Myanmar, Sri Lanka, and other places, can be seen in this light.

A social media company may be willing to bend the rules if the country is a powerful one and threatens to block its app and the company fears losing access to the market. We'll return to this point later. Most countries, though, are out of luck.

Our guess is that if social media platforms took their international commitments seriously—if they tried to pay attention to local context all around the world to the same extent that they do in the United States and Europe—they would go out of business. In other words, they are able to offer a relatively polished product with a welcoming and reasonably well-enforced set of rules in Western countries only because they have an exploitative relationship with most non-Western ones.

This section was about how AI struggles with differences across space. The next section is about how it struggles with differences over time.

AI Excels at Predicting ... the Past

There are two flavors of AI that are used in content moderation: fingerprint matching and machine learning.

Fingerprint matching is used for detecting copies of photos, videos, or audio belonging to prohibited categories that the system has previously encountered. Child sexual abuse imagery is mostly detected this way. When a social media user posts an image, it is cross-checked against databases maintained by entities such as the National Center for Missing and Exploited Children (NCMEC). This system has been unquestionably impactful in curbing a horrifying problem. Google alone sends over half a million reports of child sexual abuse imagery per year back to NCMEC, based on matches to NCMEC's database.

In this system, the work of identifying new violating images, reviewing them, and adding them to the database is largely manual. So, fingerprint matching relies on people being eternally vigilant for new child sexual abuse imagery. As in most other areas of content moderation, AI is helpful when people upload child sexual abuse imagery that was already in the database, and it's even essential for making sure reuploaded images are removed, but it won't eliminate the need for people.

In 2018, Google developed a classifier to automatically flag images as child sexual abuse. It uses machine learning rather than fingerprint matching, so it doesn't rely on the image having been encountered before. Instead, it labels images by extracting patterns, the same way a classifier can label an image as a cat even if it hasn't seen that specific image. Of course, labeling something as child sexual abuse is much more complex than labeling it as a cat, as evidenced by the system's misfire in the example of Mark and his sick child that we discussed earlier. In addition, Mark's story involves a failure of the layer of human review that is supposed to exist.

In areas other than child sexual abuse imagery, copyright, and terrorist content—hate speech, incitement to violence, self-harm, misinformation, and many others—machine learning is the predominant approach. We might hope that this cuts down the need for human involvement. Once the model learns the patterns that distinguish misinformation from accurate information, or speech that is toxic from that which is not, perhaps platforms could simply let it do its thing. Sadly, that is far from the case.

There are many ways in which patterns learned from past data are less than perfect when applied on an ongoing basis. A simple example is that language changes; new slang words enter the picture. Models need to be constantly retrained in order to keep up, and that retraining requires people labeling new posts. But this by itself isn't that much of a problem, as the human effort involved isn't too onerous.

But other types of change are much more problematic. Policies and their interpretation change constantly. Facebook's community standards have changed dozens of times since 2019. The rules that Facebook moderators use internally—which are far more detailed than the standards—change even more frequently (see figure 6.2 for an illustration of how detailed the rules are). Each of those changes requires retraining classifiers.[34]

Consider the effects of the COVID-19 pandemic. It led to a surge of misinformation, which required a great deal of manual

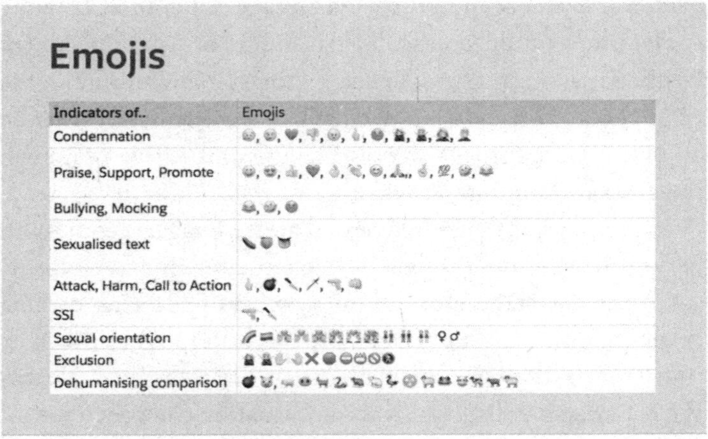

Emojis

Indicators of..	Emojis
Condemnation	
Praise, Support, Promote	
Bullying, Mocking	
Sexualised text	
Attack, Harm, Call to Action	
SSI	
Sexual orientation	
Exclusion	
Dehumanising comparison	

FIGURE 6.2. A slide from Facebook's moderator training materials in 2018. When rules are highly granular, they must change frequently to keep up.
(*Source:* Fisher M., "Inside Facebook's Secret Rulebook for Global Political Speech," *New York Times*, December 27, 2018, https://www.nytimes.com/2018/12/27/world/facebook-moderators.html.)

effort to teach classifiers about. Facebook was home to conspiracy theories ("social distancing is a ploy to install 5G towers") and dangerous health advice ("drinking bleach cures Covid"). But it took months before the company could develop classifiers for detecting these new instances of misinformation.[35]

It's critical to remember that machine learning doesn't attempt to evaluate the truth of statements. It simply relies on similarities with statements previously labeled as true or false. So, for instance, there is absolutely no way that a classifier trained before COVID would be able to evaluate statements regarding the effectiveness of COVID vaccines.

Will such a system be possible in the future using language models (the tech underlying chatbots)? It's a tempting possibility:

AI that directly evaluates the truth of statements and whose knowledge base stays up to date, either because it is regularly retrained on text from the web or because it is augmented with the ability to do web searches in real time as it evaluates a post. As of this writing, the accuracy of such a system would be far from acceptable. In a previous chapter, we described state-of-the-art language models as bullshit generators. They will likely get better in the future, but still, this type of misinformation filter would be a dangerous idea that could go wrong in unpredictable ways. Would it tolerate scientific discoveries that overturn existing consensus? Or would it be the equivalent of the seventeenth-century Catholic Church suppressing the idea that the earth goes around the sun?

Knowledge that overturns consensus is not an anomaly; it is how intellectual progress happens. So regardless of the technology behind misinformation detection, there is a risk of overreaching, resulting in a system that reinforces the political or scientific establishment while suppressing dissent. Fortunately, platforms don't try to remove all misinformation, but only certain categories of harmful misinformation. This limits the potential for overreach but doesn't eliminate it. Health information is one area where policies on harmful misinformation have been hotly contested. Besides, rather than remove content, platforms may take other actions against alleged misinformation even if it is not harmful. They may make it harder to find or ineligible for earning ad revenue for the creator. The practical impact of these penalties might be just as serious. Finally, there is always a slippery slope looming in the background of misinformation interventions: authoritarian governments often use the bogeyman of harmful misinformation as an excuse to repress speech that challenges their power.

When AI Goes Up against Human Ingenuity

We've seen that content moderation AI struggles when the content it must evaluate might not look like the past data that it trained on. There is one particularly serious way in which this might happen: people posting the content might actively try to evade content moderation.

Who might such people be? The most obvious examples that come to mind are people doing unlawful things, such as fraud, spam, or selling illegal goods. There is an endless variety of these. Cryptocurrency scammers ask victims to send crypto with the promise of quick returns on their investment, which of course never materialize. Romance scammers use fake identities with attractive pictures to entice lonely victims, then ask for money under some pretext as a condition of meeting in person. From time to time there's a panic about drug dealers using social media to lure children.

Unlike a bully or someone stoking ethnic violence, these adversaries may be technologically sophisticated. Some may even be state actors using bots to influence public opinion domestically or in other countries. Is there any hope that AI tools can resist attempts at circumvention by such entities?

This may be surprising, but we don't think law-breaking activity is one of the harder cases for content moderation. While criminals' sophistication no doubt makes life harder for platform developers, it turns out that the defenders also have important advantages that make their lives easier.

First, the professionalization of law breaking means that there is a distinct set of actors who carry out such activities—it's not like everyday people occasionally decide to try their hand at catfishing. People with criminal intent set out to perpetrate unlawful acts, which involves engaging in certain distinct patterns of communication. The line is clear, and regular users stay well clear of it. This is very different from hate speech, which actually tends to be a lot like everyday political speech—except more heated and sometimes crossing a murky line. In short, not only is the content of unlawful activity fairly distinct, so too are the networks in which the actors are embedded.

Second, platforms have a larger array of techniques at their disposal. They sometimes work with law enforcement to take down an operation or disrupt the flow of money in the underground economy. If they can identify the entities behind a scam operation and get them booted off credit card networks, the scammers lose the ability to earn money from their marks and, with it, the incentive to continue scamming. This approach can succeed even if scam detection is highly imperfect on a post-by-post level.

A historical analogy might be useful. In the early days of social media, malicious content was rampant. A computer worm called "Koobface" spread virally from friend to friend on Facebook and tricked the users into paying for fake antivirus software.[36] But soon after, platforms essentially eradicated this kind of thing. It all seems rather quaint today. While new types of unlawful activity will no doubt continue to arise, a well-resourced platform will be able to prevent it from causing major disruption.

According to the media, another area where content moderation has supposedly been kneecapped by sophisticated adversaries is state-sponsored influence operations. The evidence suggests otherwise. For example, there was widespread panic in some circles that Russia influenced the 2016 U.S. election. But research has shown that very few people were exposed to messages from Russian troll farms on social media, and that this had no discernible effect on voting patterns.[37]

Trouble arises in the occasional case where there is genuinely a gray area between scams and permitted activity. Does a user peddling a new cryptocurrency truly believe it will change the world (sadly all too common), or do they half suspect it's a Ponzi scheme and simply want to get in on the ground floor? Or consider politics. People often act in a coordinated way like bots to push a political agenda. Whether this is a political movement or an influence operation depends on how genuinely the people involved hold their beliefs. And that's a spectrum.

Ironically, the hardest kind of evasion for platforms to deal with is the unsophisticated kind that's done by everyday people. That's because platforms can't rely on evasive content coming from a distinct set of people or accounts, and they can't turn to law enforcement for help. In addition, evasion by everyday people happens on a larger scale.

Here are just a few examples of how everyday people evade content moderation:

- Overlaying the phrase "safe and effective" (a reference to a phrase frequently used by public health authorities about vaccines) on images of people who supposedly died after vaccination.
- Posting screenshots with certain words blacked out. People reading the post can understand it despite the redactions, but automated methods struggle (although perhaps the effectiveness of this method will wane as AI improves).
- "Pro-ana" communities promote anorexia, an eating disorder.[38] They use coded language like mentioning starting and goal weights. This alone is not suspicious, because the starting and goal weights are also posted by people in regular fitness communities. But users who subscribe to such content and are aware of the context can understand the intended meaning.

To appreciate how common it is for regular users to try to evade content moderation, consider *algospeak*: words or phrases that are widely understood and adopted by social media users as a way to avoid being mistakenly penalized by fickle content moderation algorithms. "Unalive" means dead. "SA" is sexual assault. "Corn" is porn, in the right context. You get the idea.

Keep in mind that algospeak is used mostly by people who are *not* posting violating content seeking to avoid error-prone content moderation tools. Ultimately, evasion strategies turn out to be very similar between policy violators (such as pro-ana communities), those who believe the policies are unjust (such as antivaccine communities), those who are systemically affected by frequent enforcement failures (Black people, trans people), and those who are affected by the crudeness of filters (everyday users). That means that it's hard to catch more policy-violating content without also affecting nonviolators.

A Matter of Life and Death

It should be clear by now that content moderation is high stakes. Nowhere are the stakes higher than when someone is about to take their own life and reveals their intention to do so on social media.

In the United States, there is one death every eleven minutes due to suicide. Worldwide, it's once every forty seconds. Other types of self-injury such as cutting oneself are even more common.

Preventing suicide has long been a goal of the medical system. But it doesn't work well: five decades of research have produced classifiers that are only slightly better than random chance.[39] The major reason, we think, is that professionals encounter patients infrequently, so the last contact with the medical system before someone takes their own life might be a few months beforehand. Predicting suicide a few months out is basically impossible. That's because suicidal ideation progresses in stages, and the final stage is reached only a few days before the attempt.[40] Before this stage, it may be clear that the patient is depressed and at risk of eventually attempting suicide, but the risk is not so serious that a major intervention is called for.

In contrast, people often post imminent suicidal plans on social media. It can be a cry for help, an attempt to find connection, or a way to record one's final thoughts. All of which means that social media is potentially a way to intervene to offer support at the moment when it is most critical.

For example, celebrities posting imminently suicidal messages on social media have often seen an outpouring of support and help from their friends and fans.[41,42] But most people posting such messages aren't celebrities and have relatively few people looking out for them. For regular people, intervention by the platform itself can make a real difference.

Facebook has employed machine learning for suicide prevention since 2017. Algorithms analyze the content of every post to score them on suicidal intent. They also analyze the comments on each post. As the company explains: "Posts that reviewers determined were serious cases of people in imminent harm tended to have comments like, 'Tell me where you are' or 'Has anyone heard from him/her?' while potentially less-urgent posts had comments more along the lines of 'Call anytime' or 'I'm here for you.'"[43] Posts that score highly are reviewed by moderators. If they confirm that there is an imminent risk, Facebook escalates it to authorities in the users' area—law enforcement or mental health professionals, depending on the country (in addition to offering resources through the app itself). To train the classifier, Facebook uses a dataset of posts previously flagged by users and labeled by moderators as indicating a risk of imminent suicide or self-injury.

Facebook does not release statistics on how often it initiates welfare checks this way. At the end of the first year of the system's operation, a Facebook employee told a reporter that the number was 3,500, but it is possible that the per-year figure has increased substantially since then.[44] To our knowledge, major platforms other than Facebook and Instagram (which are both owned by Meta) have not publicly reported having a similar system in place.

Platforms' suicide prevention efforts are valuable and important. There are philosophical questions about the morality of suicide, but as a practical matter there is consensus that people attempting suicide deserve intervention and help to improve their mental health.

At the same time, suicide prevention is a lesson in all the complicated ways that things can go wrong in content moderation. It's a bit different from the other examples in this chapter because the appropriate intervention is not simply taking down a post but rather taking some action in the real world.

No doubt the intervention saves lives in some cases, but there are also downsides. People may be subjected to warrantless searches, which might result in an arrest. Police encounters with mentally ill people often turn deadly—the risk is sixteen-fold higher.[45] Mason Marks describes many horrifying examples of such deaths that resulted from welfare checks.[46] Here's just one:

> On June 14, 2014, Jason Harrison's mother called Dallas police requesting their help transporting him to a hospital for psychiatric care. Harrison was 38 years old and had been

diagnosed with schizophrenia and bipolar disorder. When police arrived, Harrison stood in the doorway holding a small screwdriver. Despite carrying less-than-lethal weapons such as Tasers and pepper spray, two officers drew their firearms and shot and killed Harrison.

There are many other risks. Subjects of welfare checks may be involuntarily hospitalized, which can be traumatizing and dehumanizing. In many countries there are criminal penalties for suicide attempts. And people tagged as suicidal risks face difficulties getting appropriate medical care, such as painkillers, due to fears of overdosing. Finally, the intervention might actually increase suicide risk for some. That's because involuntarily hospitalized patients are often treated poorly and feel dehumanized, which might worsen their mental health.[47]

Despite suicide being so common, posts expressing suicidal intent are drops in the ocean of social media. Our estimate is that much fewer than one in a million posts express suicidal intent that then actually materializes. In other words, if the system for detecting imminent threats has a false positive rate of even one in a million, then the majority of identifications will be false positives. So, it is possible that the number of people who face the negative consequences we described is greater than the number of people who received a successful intervention.

Unfortunately but unsurprisingly, Facebook hasn't been forthcoming about how well its system works. What fraction of welfare checks were actually cases of serious imminent risk? How often were interventions too late? How many suicides did the classifier fail to catch? What are the geographic, linguistic, racial, and other kinds of disparities in the system? We doubt Facebook itself knows the answer to many of these questions.

Even if platforms told us how often their interventions were successful, it would be far from the full picture. When interventions go beyond the platform itself and enter the real world, they have second-order effects that are hard to observe, let alone predict. Knowing that it might result in a visit from the cops may make users less likely to post suicidal messages, eroding the efficacy of the system over time. Even users with no suicidal intent may be reluctant to talk about mental health online, worrying that suicide risk classifiers may be misused for other purposes like advertising or insurance discrimination.

At the other extreme, suppose tech companies became really successful at predicting suicidal intent, not just from social media posts but also from private chat messages, emails, and web searches. It is possible that public services would be tempted to cut costs by relying on those companies to refer cases for welfare checks. Yet, even if online suicide detection is more efficient and effective, it is no replacement for government services because it is not accountable to the public.

These hard moral questions will keep coming up. Continuing to deliberate them, engaging with the public, and exercising constant oversight and evaluation over the system are costs that can't be automated away.

Now Add Regulation into the Mix

In the United States, platforms have been given a free rein with content moderation under a 1996 law (and subsequent court rulings).[48] The law says that platforms can't be held liable for what people post. So platforms make content moderation decisions based on business considerations: what would make the platform appealing for users, keep costs down, and satisfy various interest groups—advertisers, social justice advocates, free speech advocates, religious groups, and countless others. That's hard enough.

If platforms did have legal liability, this risk calculus would shift massively. We know this because there is a notable exception to the 1996 law: copyright. Let's see how platforms responded to it.

YouTube is the ideal case study for copyright. Unlike most other platforms, YouTube shares ad revenue with content creators. Because of that, it has had by far the biggest copyright violation problem of all the platforms, since people try to post copyright-violating content, such as TV shows, to make money. It's on the back of this sort of unlawful activity that YouTube initially grew rapidly. Like many other tech startups at the time that took legal risks in exchange for rapid growth, YouTube ignored the problem. The company was acquired by Google in late 2006, and suddenly the risk became a pants-on-fire crisis, because Google, unlike a startup, had a lot to lose and was much more risk averse.

Under copyright law, platforms are supposed to take down content once they are informed that it infringes on copyright. Google was sued by Viacom in 2007 for one billion dollars for not doing so.[49] The case was ultimately settled out of court. What exactly platforms need to do to avoid liability still remains a bit of a gray area. Nonetheless, Google decided that it was prudent to make nice with the music and movie industries.

Enter Content ID.

Here's how Content ID works. Copyright owners can upload their content to YouTube's Content ID repository (and most major copyright owners do). When regular users post a videos, the Content ID algorithm analyzes it to see if it matches anything in the repository. It's a type of fingerprint matching, the same technology used to detect most cases of child sexual abuse imagery. It's also similar to Shazam and other apps that let you record a few seconds of music in a public place and identify the song. The algorithm is robust enough that simple modifications like flipping a video left to right or adding noise to audio aren't enough to fool it. And like Shazam, the algorithm is powerful enough to be able to identify even small snippets of video or audio that might match something in the database. Let's put a pin on that last point, because it will turn out to be important.

If Content ID finds a match between a video that's been uploaded by a user and one that's in its repository, it alerts the copyright holder, such as a music label or movie studio. YouTube allows the copyright holder to either block the video from being viewed or keep it online and claim the ad revenue from the video for itself.

We could fill this entire book just with the publicly known examples of Content ID running amok. Here are a few just to give you a taste:

- A video of a NASA rover landing on Mars—posted by NASA itself—was blocked, despite the content being in the public domain and not copyrightable.[50] The reason? News broadcasts incorporating clips from that video were uploaded and the news channels in question claimed copyright over those broadcasts, so Content ID assumed that the entirely of that content was copyrighted.
- Live performances of classical music—even of long-dead composers like Bach and Beethoven—are very often blocked.[51] Since Content ID ignores small modifications of the content, it will match two different performances of the same music. So, if another musician had performed the same piece and uploaded it as copyrighted content, it would result in a match.

- In one extreme case, a copyright claim seemed to result from a creator humming a song that contained the phrase "living on," which created a false positive match with the same phrase in an entirely different song.[52]

U.S. copyright law is actually pretty sensitive to the possibility that copyright can be abused. There is a well-developed concept called fair use which allows for satire, commentary, and many other kinds of derivative uses. But all this is far beyond the capabilities of Content ID, and so it doesn't try to account for fair uses of material. Instead, it simply finds matches and lets the two parties battle it out. Legally, users who are subjected to a false copyright takedown on an online platform have the right to issue a "counter notice," which puts the onus back on the copyright holder to sue them in court. But the prospect of entering a legal battle seems to scare off the majority of people, even when the law is completely on their side, and they don't take advantage of the counter-notice provision. A channel called WatchMojo has estimated that some two billion dollars of revenue has been stolen by copyright holders through false copyright claims.[53]

For channels that make fair use of copyrighted content regularly, such as to post movie reviews, dealing with Content ID is even more painful. Some of these channels get dozens of automated notifications (see figure 6.3 for an example). Managing all these and submitting counter notices is itself a chore. If the user doesn't get the strikes resolved speedily, it's a problem. Three strikes and you're out—YouTube removes the channel permanently.

Things have gotten quite out of hand. Scammers have been reported to send fake copyright strikes in an attempt to extort people into paying money or risk having their accounts shut

YouTube		[Copyright takedown notice] Your video has been taken down from YouTube: Top 10 Músicas do E...	7:25 AM
YouTube		[Copyright takedown notice] Your video has been taken down from YouTube: Yet Another Top 10 B...	7:22 AM
YouTube 4		Managed channel copyright strike notice - received a copyright strike. As a result, you have been is...	7:20 AM
YouTube 2		[YouTube] A copyright claim was created for content in "Yet Another Top 10 Best Lip Sync Battles" -	7:11 AM
YouTube		[Copyright claim] Your video will be taken down in 7 days: "Yet Another Top 10 Best Lip Sy...	7:09 AM
YouTube		[Copyright claim] Your video will be taken down in 7 days: "Top 10 Most DISLIKED YouTub...	7:07 AM
YouTube		[Copyright claim] Your video will be taken down in 7 days: "¡Top 10 Momentos de "A STAR ...	6:58 AM
YouTube		[Copyright claim] Your video will be taken down in 7 days: "Another Top 10 Best Eminem S...	6:57 AM
YouTube		[Copyright claim] Claim released: "Another Top 10 Songs That Will Make You Cry" - release their co...	6:43 AM
YouTube		[YouTube] A copyright claim was created for content in "Another Top 10 Songs That Will Make You...	6:42 AM
YouTube		[Copyright takedown notice] Your video has been taken down from YouTube: Top 10 DJ Khaled Co...	6:11 AM
YouTube		[Copyright takedown notice] Your video has been taken down from YouTube: Top 10 Wiz Khalifa S...	6:09 AM
YouTube		[Copyright claim] Claim released: "Top 10 Best & Worst Power Rangers Crossover Episodes" - rele...	1:10 AM

FIGURE 6.3. Channels that make fair use of copyrighted content on a regular basis must deal with an extremely high volume of bogus copyright claims.
(*Source:* Watchmojo, "Exposing Worst ContentID Abusers! #WTFU." YouTube video, May 2, 2019, 41:57, https://www.youtube.com/watch?v=Gbs9UVelEfg.)

down and all their videos deleted.[54] There have even been cases of police officers abusing Content ID to try to evade accountability.[55] When citizens filmed them, as is perfectly within their rights, these cops started playing copyrighted music on speakers, anticipating that YouTube would block the video from being uploaded.

Why has YouTube chosen this lopsided balance? Because it has far more to lose from angering the powerful interests on one side, like record labels, than small creators, who are pretty much replaceable from YouTube's perspective.

The Hard Part Is Drawing the Line

In 2016, Facebook provoked widespread outrage for removing the photograph colloquially known as "Napalm Girl" every time it was posted. It depicts a girl, then nine-year-old Phan Thi Kim Phuc, running naked in pain after being severely burned by a napalm attack. It is one of the world's most iconic photographs, historically significant for depicting the horrors of the Vietnam War and for changing public opinion about the war.

At first glance, this seems like a good example of the limitations of AI we've been talking about. Perhaps Facebook's classifiers labeled the image as child nudity, oblivious to its significance. But what actually happened was quite different, and far more interesting.

Think about it. Hundreds of millions of images are uploaded to Facebook every day; its first encounter with one of the world's most famous photographs couldn't possibly have been in 2016. Far from it. Facebook was quite familiar with the issues at hand and had used this specific image in its moderator training materials as an unacceptable one. The decision to take it down was made by people, not machines. It was not an enforcement error but a carefully considered policy choice.

More importantly, Facebook's policy here was not obviously wrong. After all, the image is so powerful precisely because it violates strongly held cultural norms against graphic bodily harm and underage nudity. As Tarleton Gillespie trenchantly puts it in his book *Custodians of the Internet*:

> This is an immensely challenging image: a vital document of history, so troubling an indictment of humanity that many feel it must be seen—and a graphic and profoundly upsetting image of a fully naked child screaming in pain.... There is no question that this image is obscenity. The question is whether it is the kind of obscenity of representation that should be kept from view, no matter how relevant, or the kind of obscenity of history that must be shown, no matter how devastating.[56]

Gillespie also points out that the photograph was released in 1972 by the Associated Press only after overcoming huge internal resistance—and at first in airbrushed form! Most newspapers that printed it had their own internal debates. Many angry letters from readers ensued. Gillespie notes that the photograph, since the beginning, has been "a proxy with which an industry and a society draws parameters of what is acceptable."[57]

The controversy that erupted in 2016 is a part of the continuing process of parameter drawing that this image evokes. It forced Facebook to back down and change its policy. In our view, the episode reflected healthy, open, necessary public debate. That's what it takes to make good policy that is seen as legitimate. It must happen over and over whenever societies need to draw lines defining acceptable speech, trading off one set of norms against another.

Social media platforms are central actors and arbitrators in these societal debates because they have set themselves up to be global public squares. They shape what is called the Overton window of speech (figure 6.4), not just through their content moderation policies but also through their design and their algorithms. Speech near the ends of the spectrum is blocked or slapped with a warning. Recommendation algorithms influence which content becomes popular. The design of the apps affects whether and how users censure others for problematic speech.

Setting the bounds of acceptable speech is a powerful way to shape public opinion and politics. Social media is often used to challenge regimes and other powerful entities, and which entities can be challenged depends on which speech is allowed. The culture wars play out on platforms, as do actual wars. And public health, and so much more. In short, social media is a "site of contestation." Interest groups of all stripes place immense pressure on platforms to change their policies.

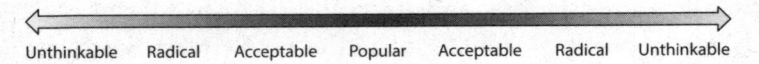

| Unthinkable | Radical | Acceptable | Popular | Acceptable | Radical | Unthinkable |

FIGURE 6.4. The Overton window of speech. Viewpoints lie on a spectrum, with more extreme ones being less acceptable.

When people complain about content moderation, they're usually complaining about policies, not enforcement—even if, as in the Napalm Girl example, they don't realize that it is the policy they have a beef with. Ultimately, this is the core reason why even the best possible AI isn't going to make people happy.

The human debate and deliberation over policies—more than the policies themselves—is the *point*. We need more of it, not less. Facebook's oversight board is a small step in the right direction.[58] The oversight board is a body that reviews important content moderation decisions by Facebook, especially those that set precedent for future decisions. It is a legally independent entity that Facebook funds. But it is made up of elites from a few countries, such as professors, journalists, and lawyers, rather than being truly representative.

Policymaking is not the only area where platforms need to be spending more effort. They also need to improve the appeals process. Recall the story of Mark and his toddler: the reason it was devastating to him is that Google refused to budge, even after the police concluded that his pictures were innocent. After this story made the press, Google introduced a new appeals process.

Google isn't the only company shutting people out of their digital lives because it won't hire enough people to rein in its AI. The problem is so widespread that having connections at tech companies is considered valuable social capital. Often, escalating an issue through an internal contact is the only way to get it looked at. One woman even claimed that she had to sleep with an Instagram employee to get her account back.[59]

At Princeton, we wanted to quantify this problem. We looked at over twenty companies. It wasn't all social media: we also looked at gig work companies like Uber, for example. In almost every case, we found a flood of online complaints: on X (formerly Twitter), on Reddit, on the Better Business Bureau website, and in app store reviews. Some of the companies we investigated had tens of thousands of complaints of unaccountable account suspension. We suspect this is the tip of the iceberg, because most people don't post their complaints publicly.

Worse than the number are the nature of the stories. We saw stories about people who lost their livelihood because they couldn't drive for DoorDash. We came across a whole company that makes apps that had to close down because Google shut down its developer account—because an *ex*-employee violated Google's policies when using his *personal* account, and Google considered the two accounts "associated."[60] (Once this story went viral on social media, Google quickly reinstated the account, as it almost always does to avoid bad PR.)

Etsy, an online market for handmade supplies, has an automated system for detecting suspicious activity. A discussion forum for Etsy sellers contains "post after post of horror stories about takedowns instigated by the AI, with no recourse, and no response from Etsy regarding the situation. Another action the AI can take is instigating 45 or 90 day holds on 75% of a seller's funds. The seller is still expected to make and ship all orders during this time. And Etsy support is non-responsive."[61]

We didn't get to hear the other side of most of these stories, and it is quite possible that any individual complaint is exaggerated or fabricated. But the fact that there are thousands upon thousands of these stories with broadly consistent patterns

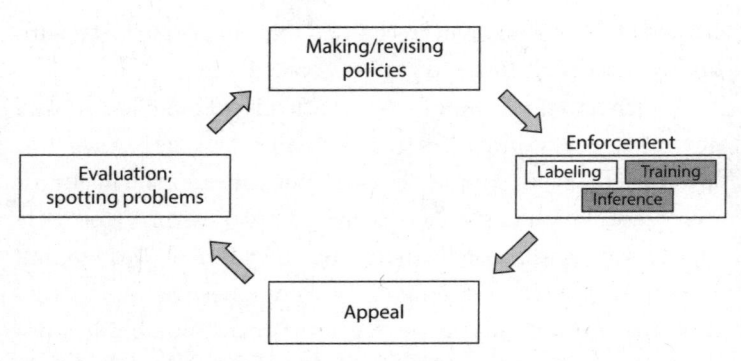

FIGURE 6.5. The stages of the AI-based content moderation cycle. Companies make and regularly revise content moderation policies. They enforce those policies through machine learning, which requires a labeled dataset of past decisions, training a model, and applying that model ("inference") on new content. Platforms must also allow appeal of decisions and must constantly stay on the lookout for problems. Only the shaded components are actually automated.

strongly suggests that there is a systemic problem—many social media and gig work companies are over-relying on AI for suspending policy-violating accounts, with little oversight or accountability.

Figure 6.5 shows one way to look at what we've discussed in this section so far: content moderation has many components, and AI has no role in most of them. (One nuance omitted in the figure is that a small part of the appeals process is also automated.[62]) What went wrong in the Napalm Girl example relates to policymaking. What has been going wrong with account suspension relates to appeals processes. Making AI better won't alleviate these problems.

There's a good argument against automating even the parts that can be automated. Human content moderators on the front lines are perhaps the best source for platforms to learn about how well their policies are or aren't working, and to spot emerging problems. Platforms currently don't seem to be making effective use of this channel of feedback, and are instead more responsive to complaints that reach the press. In 2023, they put even more distance between policymakers and frontline content moderators by outsourcing even more of the process.[63] But at least this can be changed in principle. With AI, that opportunity is cut off entirely.

What we have is a crisis of trust with multiple layers. People don't trust platform companies, not just because of content moderation but because of repeated failures on issues like privacy and antitrust. The companies are the judge, jury, and executioner when it comes to content moderation. They don't have the separation of powers that engenders some trust in government.

Worse, many of the external institutions who most strongly influence platform policies are themselves viewed unfavorably by much of the public, further undermining trust: public health agencies, advertisers, academic researchers, journalists, and governments. Public health agencies such as the Centers for Disease Control and Prevention (CDC) in the United States are weighed down by bureaucracy and failed to operate effectively during the pandemic, issuing constantly shifting and poorly communicated guidance. Social media platforms, deferring to these agencies' guidance, removed mask ads at the beginning of the pandemic and antimasking content soon after, giving people whiplash.

Governments influence platform policies through the threat of regulation or being banned. The more important a country's market is to a company, the more the company cares about keeping the government happy. And the more authoritarian a country's government is, the more likely it is to demand censorship.

These two criteria help explain why India and Turkey in particular have seen prominent clashes between governments and Western social media platforms. India's political system is categorized as a "flawed democracy" and Turkey's as a hybrid between democracy and authoritarianism. Both have been listed as among the top ten autocratizing countries of the last decade.[64,65] (While China is much more autocratic than either, Western social media apps are simply banned there.) They are both huge markets, especially India.

Facebook failed to enforce its policies against fake accounts tied to the Indian government.[66] India strong-armed both X (formerly Twitter) and YouTube into removing links to a BBC documentary critical of its prime minister. It has also set up a panel with the power to overturn social media companies' content moderation decisions. As for Turkey, its ruler got Twitter to block the accounts of some key opposing voices on the eve of presidential elections in 2023.[67]

In other countries, such as the United States, government influence is more subtle. A major flashpoint is the degree of cooperation among social media companies, governments, and academic researchers in enforcing platform policies and taking down alleged misinformation.[68,69]

In short, people have a long list of reasons to distrust content moderation processes and decisions. When platforms add AI to the mix—brittle, opaque, biased, hard to contest—the trifecta of untrustworthiness is complete.

Recap: Seven Shortcomings of AI for Content Moderation

So far, we've looked at seven reasons AI isn't going to be the solution to the problems of content moderation (see table 6.1). Most of these reasons are intrinsic, and we should expect them

WHY CAN'T AI FIX SOCIAL MEDIA? 217

TABLE 6.1. Seven reasons content moderation is hard

Cause of content moderation failures	Current AI limitations	Inherent AI limitations	Cost/business model	Inherent to social media
Context & nuance	•	?	•	•
Cultural incompetence	•		•	
The world changes over time	•	•		
Adversaries	•	?		•
Real-world consequences	•	•		•
Complying with regulation	•	•	•	
Policymaking	•	•		•

For each reason, we list whether current AI limitations contribute to why it is a problem, whether those limitations will persist for the foreseeable future, whether platforms' commercial incentives contribute to the problem, and whether the limitation is inherent to the nature of social media. ? indicates areas where there are current limitations of AI but it is unclear if they are inherent.

to continue for the foreseeable future. Some have to do with business models, and some are inherent to the nature of social media, so these factors are also unlikely to change. Let's review.

- AI is bad at context and nuance—much worse than human moderators. It is possible that this will change in the future. But a bigger problem is that companies have been unwilling to invest enough so that decisions can be both nuanced and consistent. And the fact that real-world context may not be available when evaluating a post is inevitable.
- As for cultural incompetence, there's no inherent reason a human or an automated system should perform worse in some countries than others. It comes down entirely to underinvestment.

- The fact that the world changes over time is an intrinsic barrier to the effectiveness of the AI techniques used in content moderation: machine learning and fingerprint matching.
- Turning to adversarial manipulation, even setting aside financially motivated adversaries or state actors, everyday people will continue to attempt to outsmart content moderation for a variety of reasons, some more justifiable than others. While the technical hurdles might be overcome, platforms might anger their userbase if they start punishing regular users for evasive behavior.
- When problematic content requires a platform to take some real-world action, a new set of challenges arises because it is hard to predict in advance what the real-world effects of those actions will be.
- Regulation tends to result in "collateral censorship," where platforms moderate more aggressively than they need to so as to minimize legal risk to themselves. Avoiding this would require platforms to actually evaluate the lawfulness of contested content, which runs into both technological and cost barriers.
- Policymaking will remain an extremely challenging aspect of content moderation because social media is a site of political contestation. AI has no role in this essentially human activity.

A Problem of Their Own Making

The real tragedy of platforms' failures with content moderation is that the problem itself is largely of their own making. Hate speech is rampant on Facebook because Facebook incentivizes it. Thousands of children have participated in life-threatening trends like the "blackout challenge" on TikTok (where people try to hold their breath long enough to pass out) because TikTok's algorithms reward them with attention. And yes, challenges like this have resulted in the deaths of children.

We're not saying that engineers programmed algorithms to amplify hateful and dangerous content. But they've made a series of deliberate design decisions that cumulatively have had this effect.*

First, social media algorithms optimize for engagement: how often we click, like, comment on, or otherwise interact with posts. But the content we impulsively interact with is not what brings us value. Platforms have no good way to measure what matters. So they end up optimizing for the wrong thing.[70]

The gap between what we actually want and what algorithms decide we want seems to be getting bigger in some ways. On TikTok, the algorithm's logic is based on how often users stay on a video before scrolling. That means it caters to our basest impulses: there's so much trashy content we'll linger over but won't explicitly engage with—a kind of digital rubbernecking— and this kind of content is rewarded when the algorithm focuses on our unconscious, automatic signals. TikTok's design has been so commercially successful at getting users to spend more time on the app that other platforms are desperately trying to copy it.

Engagement optimization incentivizes people to post polarizing content. If a post is cheered on by one group while another group silently disapproves, the algorithms only see the positive reaction. Worse, if the disapproving group expresses

---

* Measuring the effects of platform decisions on societal outcomes is challenging. It is a debated and ongoing area of research. This section represents our best interpretation of the available evidence.

their anger instead of being silent, that is usually counted as increased engagement and only fans the flames.

Researchers have published algorithms such as bridging-based rankings where content that brings people together is rewarded over content that pushes them apart.[71] So far, no major platform has adopted such ideas. The only one to have tried is X (formerly Twitter). In 2022, the company launched Community Notes, a feature that allows users to correct misleading or incorrect information in posts.[72,73] In an attempt to bridge the gap between people with differing political beliefs, notes that are accepted by people across the political spectrum are more likely to show up on posts. But the overall impact of Community Notes on the platform is small, since only a small fraction of posts are egregious enough to receive a note. And critically, the feature doesn't change the algorithm that determines which posts show up on feeds, which continues to amplify divisive content.[74]

Mark Zuckerberg once observed that content that is closer to violating content moderation policies is more likely to be engaged with. For example, harmful misinformation is usually prohibited, so content that misinforms without directly being harmful would fall into this category. He then called this phenomenon the "natural engagement pattern" (figure 6.6).[75]

Calling the amplification of problematic content "natural" is a disingenuous move that deflects the platform's responsibility. There is nothing natural about it—social media is a highly artificial environment where people adjust their behavior based on what they think will be rewarded. So, whatever engagement patterns arise are heavily influenced by design and algorithms.

There is a curious fact about this tug of war in which recommendation algorithms amplify harmful content and content moderation algorithms try to detect and suppress it: the

Natural engagement pattern

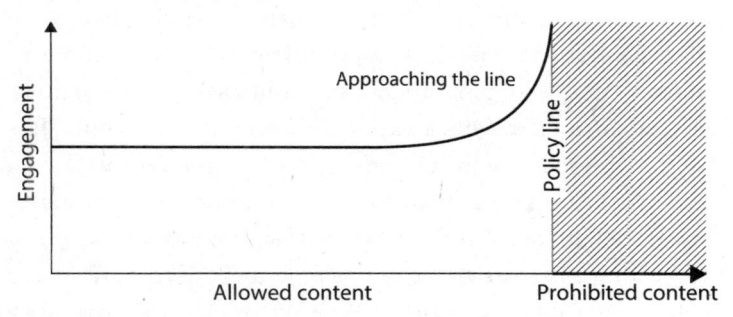

[Graph showing engagement on y-axis, with "Allowed content" to "Prohibited content" on x-axis. Curve labeled "Approaching the line" rises sharply near the "Policy line" separating allowed from prohibited (hatched) region.]

FIGURE 6.6. According to Mark Zuckerberg, the amplification of problematic content is a natural phenomenon. We disagree. (Redrawn from Zuckerberg, "A Blueprint for Content Governance and Enforcement," last updated May 5, 2021, https://www.facebook.com/notes/751449002072082/.)

recommendation algorithms have the upper hand. Earlier we described how people stay one step ahead of algorithms. For example, pro-ana posters use coded language like talking about starting and goal weights. You might think that when people make posts invisible to content moderation algorithms, they also become invisible to recommendation algorithms. But that's not the case![38] The reason is that recommendation algorithms primarily operate not based on the contents of posts, but based on the behavior of people when encountering that content. As long as people can correctly decode the meaning of a post, recommendation algorithms will pick up on it. The logic is something like this: "Lots of people who previously liked pro-ana posts seem to like this one, so this post should be shown to more people who have liked pro-ana posts in the past."*

---

* Why can't content moderation algorithms also use this network-based logic? Companies have tried, but there are subtle technical reasons this is much harder.

There are many types of harmful content that get amplified by recommendation algorithms but can't be dealt with by banning. For example, some harmful content exploits the gray area inherent in any content moderation policy. As social psychologist and former Facebook executive Ravi Iyer points out: "It is often more effective to misinform people implicitly by taking a true event (e.g., an adverse reaction to a vaccine, a rare instance of voter fraud, or a crime committed by a minority group) and suggesting it is representative of a broader pattern, than it is to make a verifiably false claim."[76] Some content is not harmful by itself but because of the effect it generates in some (but not other) people who see it. Social media has had a devastating effect on teens' mental health by encouraging social comparison, but good luck banning Instagram selfies. And as for hate speech, it is often more effective to incite violence against a group by spreading fear about that group than by calling for violence directly.[77]

Another limitation of content moderation, which operates on a post-by-post basis, is that some content becomes harmful only in the aggregate. There have been reports of prominent tourist and religious sites being ruined by a horde of influencers, each looking to score views with their Instagram or TikTok videos.[78] The impact of any one tourist is minimal, but the competition among them results in a tragedy of the commons. Another example: a single individual calling out another for a (real or perceived) social transgression is an everyday occurrence, but when thousands of people dunk on one target, as often happens on social media, it can ruin lives and careers. Recall the story of Justine Sacco from chapter 3—the woman whose life fell apart after her attempt at humor about AIDS in Africa was misunderstood and went viral. Finally, consider polarization. Individual instances of divisive or polarizing rhetoric aren't directly harmful, and users would rebel if every divisive post were taken down. But in the aggregate, they can drive societies apart.

Iyer is blunt about his assessment: "Content moderation is a dead end."[75] The problems with social media are inherent in their design and cannot be fixed by the whack-a-mole approach of content moderation. We agree.

The Future of Content Moderation

We can't predict the future. And given the themes of this book, we would be especially foolish to try. But we can certainly talk about what is and isn't possible when it comes to content moderation, what is likely, what would be desirable, and what some trends are that are already in motion.

We can be fairly confident that there will continue to be improvement in the "easy" part of content moderation AI— achieving parity with human content moderators. But all the hard parts will continue to be hard. These are the issues we discussed that go beyond the accuracy of individual decisions: investing in understanding local context, dealing with constant social change, managing users' attempts to evade content moderation, navigating the highly political process of making policies for hundreds of countries and cultures, and complying with regulations.

Speaking of regulation, the EU's Digital Services Act came into force in 2023. It's a wide-ranging regulation that includes provisions relating to content moderation. It requires more transparency about automated content moderation tools and better oversight of moderation decisions. These requirements are well intentioned, and we agree that more transparency is needed as well as a better appeals process for high-impact decisions such as account deletion. But most decisions don't rise to that level, and billions of removal decisions are made every year. Daphne Keller, former Google general counsel and now a scholar at Stanford, argues that by making moderation so expensive, the law will force platforms to moderate less.[79] In our view, policymakers should recognize that flawed individual decisions made using AI aren't the problem so much as a symptom of deeper, underlying structural issues.

Along similar lines, Evelyn Douek, another Stanford scholar, says we should look at content moderation through the lens of "systems thinking," which puts the focus not on individual decisions but on the broken processes that produce them.[80] Systems thinking leads to interesting recommendations for reform. For example, Douek argues that "platforms should be required to put a wall between those concerned with the enforcement of content moderation rules, on the one hand, and those whose job performance is measured against other metrics, such as product growth and political lobbying, on the other. This should be enforced through fines if the latter interfere with the decision-making of the former regarding individual content moderation decisions." Regardless of whether a government-mandated separation of functions is the right answer, we agree that improving content moderation will require institutional reform.

Reforming trillion dollar companies won't be easy, and the public square is too important to be left in this broken state. Are there alternatives outside of the Big Tech companies? One model worth looking at is Reddit. There, rather than one giant network of users, discussion is broken up into topic-focused "subreddits." The critical difference is that moderation is done by volunteer members of those subreddits rather than company employees or contractors.[81] They are assisted by automated tools developed centrally by the platform.

This model has many admirable qualities. Since moderators are members of the specific community that a post appears in, they are keenly aware of context rather than oblivious to it. It gives users a voice. If they don't like the moderation policies or enforcement, they can talk to the moderators or perhaps consider becoming a moderator. Furthermore, the Reddit model has the separation of functions that Douek recommends, with moderation being relatively independent from the company's commercial side.

Each subreddit is free to set its own moderation policy. This leads to healthy experimentation with content moderation, which has even been the subject of study by researchers and has led to a lot of knowledge about what works and what doesn't. It also gives users choice rather than forcing everyone regardless of culture or personal preferences to accept one set of policies. On the other hand, it means that many problematic communities find their home on Reddit.

Finally, it's unclear if Reddit-style moderation can scale to sites as widely used as Facebook or YouTube. Reddit's automated tools are relatively rudimentary, perhaps because of the lack of standardization of policies. And for all the positives of volunteer content moderation, it is certainly less efficient than the factory model. The amount of moderation labor on Reddit appears to be several orders of magnitude lower than on Big Tech platforms.[82] This would not be sufficient to enforce a stricter set of policies such as Facebook's.

An even more radically decentralized model is seen in Mastodon. On Mastodon, the user experience is roughly like X (formerly Twitter), but it works quite differently under the hood. Users join a specific Mastodon server, of which there are thousands. The servers all talk to each other, so users can follow others on any server, but content moderation policy and enforcement is on a per-server basis.

The decentralization of moderation combined with the unification of content results in an inescapable tension: a server has no recourse against users who violate its policies if they are on another server—except to block that server outright (blocking individual users doesn't scale). Subreddits don't have this problem because communities are distinct. And the traditional platforms don't have this problem because moderation is centralized. In many ways, Mastodon's content moderation is the worst of both worlds.

In short, the bad news is that there is currently no obvious alternative to the mainstream platforms and their way of doing things. The good news is that Big Tech's hold on social media seems to be loosening, so there's a lot of experimentation happening.

If there's one thing we want you to take away from this chapter, it's this: content moderation is another example of the fact that failures and limitations of AI have less to do with AI and more to do with the institution adopting it. Social media platforms have set themselves up to be vehicles of entertainment, tools for social connection, and global public squares, all in one. It has been a rocky experience so far, and it is far from obvious whether rolling these three into one can be done in a responsible way.

Increasing the adoption of AI for content moderation can make the process a bit more efficient, but it does nothing to alter the deep-seated contradictions inherent in the concept. When content moderation AI is portrayed as a solution to the moral and political dilemmas of social media rather than merely a way for companies to save money, it becomes a form of snake oil.

Chapter 7

WHY DO MYTHS ABOUT AI PERSIST?

SEPSIS IS DEADLY. It is an immune system response to infections, and it can lead to tissue damage, organ failure, and death. It is a leading cause of death in U.S. hospitals and causes one in every five deaths globally.[1] Early detection can prevent deaths: the earlier sepsis is detected, the easier it is to treat.

Many companies have claimed that hospitals can use AI to predict the risk of sepsis using electronic health records. These records store detailed information about each patient, including their medical history, test results, and the medications they are currently using.

Epic, a U.S.-based healthcare company, has the country's largest collection of health records, with information on over 250 million people in the United States.[2] In 2017, armed with this vast amount of data, Epic released an AI product to detect sepsis. It was a plug-and-play tool that hospitals could use with their existing health records. The value proposition was clear: hospitals could decrease deaths due to sepsis without spending more money on equipment or data collection.

Epic was not shy about showing off its adoption rates. Hundreds of hospitals had adopted the system, and Epic claimed its model reduced mortality rates due to sepsis in these hospitals. The model was hailed for allowing clinicians to spend more time with patients.[3] In a 2020 interview, Epic CEO Judith Faulkner said: "If you put in sepsis AI . . . six hours before the human being can tell that this patient is coming up with sepsis, it can identify [sepsis] in many cases and save lives."[4]

Epic didn't release any peer-reviewed evidence about the model's accuracy. Like many other AI companies, it said its model was a proprietary trade secret. External researchers could not verify Epic's findings. Four years after its release, no third-party evaluations of the model's efficacy existed, even as hospitals continued adopting it.

Finally, in June 2021, researchers from the University of Michigan Medical School released the first independent study of the model.[5] The researchers could perform the study only because they worked at a hospital that used it. They had hospital records of patients about whom the model had made predictions. They also had data to check if the patient eventually suffered from sepsis.

The results were shocking. Epic had claimed that its model had a relative accuracy between 76 percent and 83 percent. (Relative accuracy refers to the probability that a patient who would go on to develop sepsis would be rated as higher risk than a patient who wouldn't.) But the study found that the relative accuracy was actually 63 percent—far worse than originally claimed. A relative accuracy of 50 percent means it is as good as flipping a coin. So a 63 percent relative accuracy means that the model is only slightly better than random.

Epic responded to the damning study by highlighting anecdotal evidence from two organizations that claimed that the model had improved mortality rates.[6] The company also noted that tens of thousands of clinicians used the sepsis model. Surely if so many hospitals used the model, things couldn't be too bad?

It turned out that while Epic was flaunting its high adoption rates, the company was simultaneously paying hospitals up to USD 1 million in credits if they satisfied certain conditions.[7,8] One of the conditions was the use of a sepsis prediction model. It is unclear if hospitals used the model because it worked well or because it helped their bottom line.

In October 2022, Epic stopped selling its one-size-fits-all sepsis prediction model. Instead, it asked hospitals to train the model on their own patient data before using it for sepsis prediction.[9] After years of insisting that a plug-and-play model could save lives, Epic had walked back on its claims. A big selling point of Epic's models was that they did not require any extra hospital investment. They could be used across hospitals out of the box, using existing health records. But if hospitals need to create AI models locally, they lose many of the efficiency gains that plug-and-play AI promises.

The sepsis prediction model was one of many models released by the company. In some of their other models, Epic included features such as "religion"—for instance, to predict which patients won't show up for their medical appointments, likely leading to discrimination based on religion.[10] Though the company rolled out a fix once this misstep was made public, this is another example of the inadequate scrutiny on AI products sold to hospitals.

Epic's sepsis model is a cautionary example of the AI hype cycle. We have seen this story play out countless times: A company releases a new AI application with much hype but does not disclose how it is trained or what data it uses. Journalists repeat the company's claims, quote spokespeople, and build on the hype. Even though there is no public evidence that the tool works well, it is quickly adopted in consequential settings based on unverified claims. In many cases, its use continues unchallenged. And when it finally comes under scrutiny, researchers and journalists working to expose its flaws face an uphill battle. In Epic's case, it took a years-long academic study and relentless reporting by STAT, a health-oriented news website, to get the company to fix its models.

Over the course of this book, we have seen that people believe all sorts of myths about AI, and that those myths can help prop up AI snake oil. Why do myths about AI persist? This chapter sketches out an answer. We will go over the major sources of hype: companies, researchers, journalists, and public figures, and see how they exploit cognitive biases to misinform the public.

Companies have commercial interests in spreading hype about AI—they want to sell more of their products. And so they talk up the impact of AI in "revolutionizing" their industry. Investors like to fund groundbreaking AI, so in some cases, companies hype their "AI," even when it is humans pulling the strings behind the scenes.[11] Calendar scheduling company x.ai (not the same as Elon Musk's recently launched AI company) advertised that its AI personal assistant could schedule meetings automatically, claiming, "Our scheduling AI will send time options to your guests taking into account any additional details from you."[12,13] In fact, the company tasked humans with reading and correcting errors in nearly every email generated by its AI scheduler. Live Time claimed to use AI to detect events that could harm public safety, such as active shooters. It raised over USD 200 million and secured a USD 20 million contract with the U.S. state of Utah. But an audit of Live Time found that the company did not use AI at all.[14]

Meanwhile, research in AI is facing a crisis of reproducibility. A lot of AI research is not independently verified, and without oversight, researchers have incentives to exaggerate the impact of their findings to garner attention and funding. Even if they act in good faith, it is easy to make errors that overestimate the performance of AI applications.

Chronically underfunded news outlets amplify hyped claims by researchers and companies. Journalists often do not have the time or expertise to verify these claims comprehensively, so they churn out lightly edited PR statements as news. And they amplify claims made by public figures, such as public intellectuals and tech CEOs, rather than engaging in grounded discourse.

In addition, humans share cognitive biases that make us especially susceptible to hype. For example, we anthropomorphize AI—we treat it as if it is an agent with humanlike qualities. That leads to misplaced trust in AI systems.[15] Cognitive biases can also prevent us from recognizing the limits of our own knowledge. We are often overconfident in our understanding of how things work—we feel that we understand complex phenomena in more detail than we actually do.[16]

In the midst of this hype, it is hard for any of us to evaluate claims about AI critically. This chapter offers a remedy. Understanding how myths about AI are produced can help us build up resistance to snake oil.

AI Hype Is Different from Previous Technology Hype

The best known way to track hype in technology is the Gartner hype cycle.[17] Created by the consulting firm Gartner, the hype cycle depicts the five stages that emerging technologies go through in their life cycle.

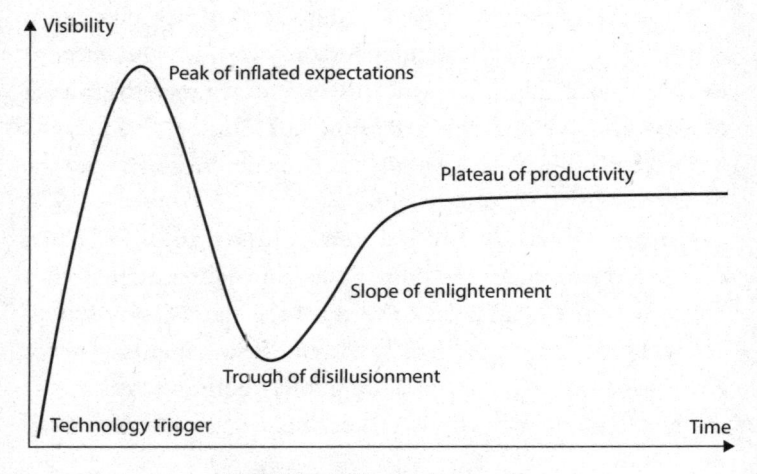

FIGURE 7.1. The Gartner hype cycle.
(*Source:* Jeremy Kemp at English Wikipedia, CC BY-SA 3.0.
https://commons.wikimedia.org/w/index.php?curid=10547051.)

The Gartner hype cycle, shown in figure 7.1, provides a commonsense analysis of hype: when a new technology is created, it sharply rises in visibility to the "peak of inflated expectations" when public visibility of the technology is the highest. Invariably, these expectations are not met, leading to the "trough of disillusionment." After the product proceeds along the "slope of enlightenment," when productive real-world uses are found, it finally settles into the "plateau of productivity" when it gains mainstream adoption and success.

Where does AI technology belong on this curve? In 1995, when the original curve was published by Gartner, "intelligent agents" were already at the peak of inflated expectations. Are we now in the trough of disillusionment? Or have technologies like ChatGPT lifted AI into the plateau of productivity? In theory, if we could locate AI at the peak of inflated expectations, we could accordingly temper our optimism until the tech is ready for real-world adoption. Or if we realize that AI is actually in the trough of disillusionment, it might make sense to plow ahead despite the issues.

Unfortunately, the Gartner hype cycle is not a good way to track the adoption and usefulness of AI. Technologies rarely evolve per the hype cycle.[18] Every year, Gartner releases a list of technologies alongside their place on the hype cycle. Over a quarter appeared on the hype cycle for only a single year. Only a handful went through all the stages to enter mainstream success. Some technologies die out quickly, while others take decades longer than expected. But the hype cycle has no way to account for unsuccessful technologies—there is no "failure" stage. There are successful technologies that indeed follow the hype cycle, like the World Wide Web with its dot-com bubble and bust. While we remember technologies that follow this pattern and survive, we don't typically think about technologies that never became successful or useful in the real world.

The inadequacy of the Gartner hype cycle is even more sharply apparent when it comes to AI, because AI is an overarching term for many different technologies. Some types of AI have fundamental limits, such as those that claim to predict people's future. So even if the Gartner hype cycle were useful, different types of AI would be at different stages of it. In fact, AI research has historically seesawed between peaks and troughs instead of following the Gartner model.

As another example of how AI hype differs from other tech hype, let's compare AI to cryptocurrency. Crypto started with the release of Bitcoin in 2009, but many applications have been proposed for the underlying technology, including decentralized art ownership and social media. These are collectively called Web3.

There are similarities between AI and Web3. Both are umbrella terms. Both have been heavily funded by venture capitalists. Like AI companies, Web3 companies have generated hype and haven't always been truthful in their claims. Like AI, the harms of crypto applications are enormous. Bitcoin mining alone consumes more energy than entire countries such as Denmark, Chile, or Finland.[19]

The hype around crypto reached a fever pitch in early 2022. Crypto exchanges—platforms where you could buy or sell cryptocurrencies—spent hundreds of millions of dollars on advertising campaigns. Celebrities like Larry David and Matt Damon appeared in commercials during the Super Bowl. In June 2022, four months after the Super Bowl, the value of Bitcoin had dropped more than 50 percent. In November 2022, FTX, the third-largest crypto exchange, went bankrupt. Customers lost over USD 11 billion. The CEO, Sam Bankman-Fried, often seen as the poster boy for crypto, was convicted of fraud, conspiracy, and money laundering. The celebrities who promoted FTX also faced a class action lawsuit.[20] During the 2023 Super Bowl, a year after the glitzy ads, zero crypto ads aired.[21] Web3 critic Molly White documents the monetary harms of crypto hacks and scams.[22] Since cryptocurrencies are for the most part unregulated, victims of scams have little recourse. Between 2021 and 2023, over USD 50 billion was lost to such scams.

Is AI headed for a similar crash?

There is one fundamental difference between AI and crypto. Despite being touted as the future of the internet, crypto and Web3 lack socially beneficial uses. This is not empty cynicism; it is an insight born from experience. In 2016, Arvind coauthored a textbook about the computer science of Bitcoin and cryptocurrencies. The book has been used in over 150 courses across the world, and an online course based on the book has had over seven hundred thousand enrolled students. When Arvind began working on the book in 2014, the tech was still new, and it seemed plausible that people would build useful products using it. But it has gradually become clear that crypto is a solution looking for a problem. Since 2018, Arvind has not worked on developing cryptocurrency technology and his main interest in this area has been helping shape public policy to counter its harms.

In comparison, AI can be extremely useful. Most apps on our phones use AI in some form. The trouble with AI hype stems from the mismatch between claims and reality. While "just don't believe it" is adequate advice for countering crypto hype, resisting AI hype requires a more nuanced approach.

The AI Community Has a Culture and History of Hype

Contrary to the Gartner hype cycle, the field of AI has a history of cycling between peaks and valleys. The peaks are called springs—periods of intense growth, funding, and hype. The valleys are called winters—periods where funding dries out, research stagnates, and expectations deflate.

As we discussed in chapter 4, in 1958, Frank Rosenblatt demonstrated that a machine learning algorithm called the perceptron could classify images. The result was widely celebrated at the time, but a decade later, MIT researchers Marvin Minsky and Seymour Papert showed that perceptrons could solve only certain limited problems. And in 1972, a critical report by mathematician James Lighthill, which was commissioned by the UK government, said that much of the progress toward building general AI systems was illusory.[23] These were critical blows that resulted in funding for the field drying up and precipitated the first AI winter.[24]

In the 1980s, research in AI heated up again. This time, it was the rise of so-called expert systems that prompted the excitement. Researchers would interview domain experts, such as doctors, to formulate rules and heuristics based on how those experts actually made decisions. Expert systems then used these rules to make their own decisions. While they were useful in some settings, expert systems were brittle, as they could not be used effectively in situations for which no rules existed. It was also hard to update these systems based on new information. As a result, by the end of the decade, the hype and funding had all but disappeared. "When I received my PhD in 1990," writes noted AI researcher Melanie Mitchell, "I was advised not to use the term 'artificial intelligence' on job applications."[22]

The alternating winters and springs show that the history of AI is littered with overoptimism about its capabilities and utility. In the short term, hype can attract massive investment and lead to intense growth. But this hype also sets a high bar for real-world impact. AI winters result when the usefulness of AI applications doesn't live up to the hype.

AI research relies on corporate funding.[25,26] Modern AI technologies, such as chatbots, incur massive costs to develop, both toward hardware and researcher time. This investment is untenable for most academic groups. Instead, corporations such as OpenAI, Google, and Meta, formerly Facebook, have developed the most powerful recent large AI models. AI researchers increasingly follow the money, forming collaborations with corporations instead of working independently. Almost three-quarters of AI PhDs choose to accept industry positions instead of entering academia, up from just a quarter two decades ago.[27]

Observations about computer science research aiding industry applications have been made for decades.[28] But most academic computer scientists don't consider the cozy relationship with industry to be problematic. It is considered perfectly acceptable to have the field's research interests defined by applicability in the industry. One side effect of this long-standing relationship with industry is that academic research is of limited effectiveness as a check on industry power.

In this regard, AI stands in contrast to fields like medicine. Here too, the influence of corporate funding is clear.[29] For instance, pharmaceutical companies routinely fund research pertaining to products or drugs they themselves produce, leading to questions about the quality of scholarship. But in this case, the corrupting influence of industry-funded research has led to big debates in the field, is the topic of several books, and has led to strong norms and rules on conflict-of-interest disclosure.[30,31]

Another reason for hype in the AI community is a lack of focus on scientific understanding. Instead of scientific explanations for *why* AI works well, the community focuses primarily on improving the performance of AI on benchmark datasets. This makes sense given the funding and influence from the industry. Corporations value engineering breakthroughs (that can be incorporated into profit-making products) more than scientific understanding. Companies often launch products before they (or academics) can explain how the products work, so new tech can feel like magic. As a result, researchers understand which AI techniques work well. But we don't understand *why* they work well, because of the lack of time and resources devoted to this topic.

This trend has not gone unnoticed. In 2017, AI researchers Ali Rahimi and Benjamin Recht won the Test of Time Award at NeurIPS, one of the world's largest AI conferences. A Test of Time Award is awarded to papers that are ten years old and lead to significant impact on the field. After the award was announced,

Rahimi was invited to give a speech. His twenty-minute speech was a scathing indictment of AI research,[32] which he compared to alchemy due to the lack of rigor and low standards of evidence. Rahimi criticized the field's focus on beating previous results on benchmark datasets. Instead, he asked researchers to understand why AI tools perform well. "If you're building photo-sharing systems, alchemy is okay. But we're beyond that now. We're building systems that govern healthcare and mediate our civic dialogue," he said. "I would like to live in a society whose systems are built on top of verifiable rigorous thorough knowledge and not on alchemy."

Rahimi isn't alone in his criticism of the community's culture. "In 1892, the psychologist William James said of psychology at the time, 'This is no science; it is only the hope of a science,'" writes Melanie Mitchell. "This is a perfect characterization of today's AI."[22] In 2018, researchers Zachary Lipton and Jacob Steinhardt wrote a paper titled "Troubling Trends in Machine Learning Scholarship," in which they highlighted several recurring problems with AI research.[33] For instance, researchers often make speculative claims about AI that, because of the credibility of the authors, are assumed to be true without empirical evidence. Researchers also misuse language to imply that AI tools perform better than they actually do—for instance, by implying that they have human-level reading comprehension, when the only evidence is on a benchmark dataset instead of evaluations in the real world.

This culture is exemplified by a dismissive attitude toward domain experts that many AI researchers and developers hold. In 2016, AI pioneer Geoffrey Hinton claimed: "If you work as a radiologist, you're like the coyote that's already over the edge of the cliff but hasn't yet looked down, so doesn't realize there's no ground underneath him. People should stop training radiologists now. It's just completely obvious that within five years, deep learning is going to do better than radiologists."[34] In 2022, there was a worldwide shortage of radiologists.[35] AI has not even come close to replacing radiologists.

Companies Have Few Incentives for Transparency

Let's go back to Epic's sepsis model. Epic never released it publicly, so it faced no external scrutiny. Unlike peer-reviewed research, no independent reviewers verified the claims. Epic is not an outlier; whether for recidivism-prediction models such as COMPAS or for job-candidate-assessment models such as those built by HireVue, companies don't make their models publicly available for scrutiny, arguing that they are trade secrets.

It is no surprise that when AI companies have skin in the game, they put their business interests above transparency. Across industries, companies with economic incentives to hide shortcomings will do so. For instance, it took decades to establish the link between smoking and cancer. Big tobacco companies lobbied researchers, distorted early studies that linked smoking and cancer, and spent millions of dollars trying to falsely imply that there is no long-term damage to the lungs caused by smoking. Fossil fuel companies Shell and Exxon knew about the climate costs of their products in the 1980s.[36] They never made this knowledge public. Instead, they actively downplayed the harms and lobbied to prevent legislation that could address climate change.[37,38]

In 2021, our colleagues Amy Winecoff and Elizabeth Watkins conducted a study of early-stage AI startups.[39] They interviewed twenty-three entrepreneurs to understand how startups use AI. Unsurprisingly, they found that because investors want to invest in applications with high accuracy, companies game the metrics to report high accuracy numbers.

Here's one example. A popular way to measure accuracy for image classification is "top-N" accuracy. When an AI model tries to label an image—say, a dog photo—it outputs many possible guesses. Top-N accuracy asks if the model can guess what's in the image in its first N tries. If N is three, and the model's first three guesses are cat, dog, and lion, we would give the model a perfect score since the correct label (dog) is one of its first three guesses. Increasing the number of tries makes the task easier. The top-5 accuracy of a model will always be better than the top-3 accuracy because the model has five tries to label the image correctly. Similarly, top-10 accuracy will be even better.

When describing how startups measure accuracy, one developer said:

> How it's measured is we have to make sure it's 90% or above [...]. So if we need to switch from top-3 accuracy to top-5, just people seeing a 9, they don't even think about what it's measuring.... People just have artificial concepts of what's good and what's bad.[40]

In other words, developers increase N until the accuracy hits 90 percent. At that point, the company becomes attractive to investors, even if it performs poorly otherwise. Companies have many other similar ways to fudge accuracy numbers and make their products look better than they are.

It's not just entrepreneurs who fudge numbers. The venture capitalists (VCs) funding them also have the same incentive:

> The VCs wanted to hype things up, get a lot of press, make a splash, so they could raise the next round at a higher valuation and look good to their [partners], which was actually contrary to what we needed to do for the slow growth to build the business.

Even if companies do not fudge accuracy measurements, performance on benchmark datasets overestimates the usefulness of AI in the real world. As we saw in chapter 4, the dominant way to determine the usefulness of AI is through benchmark datasets. But benchmarks are wildly overused in AI.[41] They have been heavily criticized for collapsing a multidimensional evaluation into a single number.[42] When used as a way to compare humans and bots, the results can mislead people into believing that AI is close to replacing humans.

For example, OpenAI claimed that "GPT-4 exhibits human-level performance on the majority of these professional and academic exams," and that GPT-4 scored in the ninetieth percentile on the bar exam.[43] Many took this as a sign that AI will soon be good enough to replace lawyers. But a lawyer's job is not to answer bar exam questions all day. Real-world utility is different from good performance on a benchmark. Moreover, professional exams, especially the bar exam, notoriously overemphasize subject-matter knowledge and underemphasize real-world skills, which are far harder to measure using a standardized test.[44] So not only do these exams fail to capture the real-world utility of AI, they also overemphasize precisely the thing that AI is good at.

The Reproducibility Crisis in AI Research

AI companies aren't the only parties motivated to spread hype. Many of the advances you hear about in the news come from researchers. But these advances are tenuous, as AI research is suffering from a reproducibility crisis. What is reproducibility? And why is it important?

Imagine a world where you get a different result every time you run a scientific experiment. It doesn't matter how carefully you set up your instruments or how meticulously you measure the outcomes of interest. In this world, we cannot measure Earth's gravitational pull, so we cannot calculate the trajectory of an object flying in the air, and ultimately cannot come up with the science that enables airplanes or moon landings. We cannot test medicines and vaccines to see if they are reliable, so we cannot rely on new treatments for deadly diseases and cannot control or prevent the worst effects of pandemics.

Reproducibility, or the ability to independently verify the results of a scientific experiment, is a key component of scientific research. If scientists cannot run the experiments in a study multiple times with the same results, they cannot trust the results.

How do we verify if the results of a study hold up? We often do not have the resources to run each experiment multiple times. Experiments can be costly or can require a long time investment from researchers to run properly. Further, not all scientists have access to the tools required to run each experiment. Today, the main way to evaluate the quality of scientific research is through peer review. When researchers release a paper, experts in the field—usually between two and five researchers who have previously worked on similar topics—will assess how rigorous the study is. Peer review is a coveted stamp of approval on a research study. But peer review is not a panacea, and errors can still creep in.

When scientific fields have tried to test reproducibility systematically, they have found that many peer-reviewed studies fail to reproduce. Perhaps the most prominent example comes from psychology. In 2015, a large group of researchers tried to replicate past research in social psychology. They found that only 36 percent of published results could be replicated, despite being peer reviewed and published in top scientific journals.

When other fields have tried to replicate past findings, they have found similar results: many studies fail to reproduce.[45,46]

In a 2018 study, Odd Erik Gundersen and Sigbjørn Kjensmo from the Norwegian University of Science and Technology set out to investigate the reproducibility of AI research. They reviewed four hundred papers from leading AI publications to ascertain if they contain enough detail to be reproducible by an independent researcher. They found that *none* of the four hundred papers satisfied all of the criteria (such as sharing their code and data) for reproducibility. Most papers satisfied merely twenty to thirty percent of the reproducibility requirements they identified, making it hard to even investigate if the results were reproducible.

Over the last few years, a major focus of our own research has been on reproducibility. In 2020, during a graduate seminar on the limits to prediction, we looked at which outcomes are predictable and which are not. We found that prediction is hard for the vast majority of social outcomes, such as those involving people. This echoes what we saw in chapter 3. The only outlier seemed to be civil war prediction, a subfield of political science that aims to predict which countries and regions will experience civil wars in a given period. As you can imagine, predicting political conflict and violence in advance is extremely hard. But in 2016, a paper published in a top political science journal claimed to predict civil wars with astounding accuracy using AI.[47] Since that 2016 paper, several others claimed that AI can far outperform older statistical techniques for predicting civil war.[48,49] We were curious to find out what allows AI to perform so well for civil war prediction when it performs so poorly for other social outcomes, so we decided to dig deeper.

To our surprise, we discovered an error that resulted in overoptimism about the performance of AI. The AI models were evaluated on data they had already been trained on—teaching to the test. The error is known as leakage, and it violates a cardinal rule of AI: never test on the training data. When we fixed the error, AI performed no better than decades-old models.

Leakage is well known in AI. In an apocryphal story from the early days of computer vision, a classifier was trained to discriminate between images of Russian and American tanks with seemingly high accuracy. It turned out, however, that the Russian tanks had been photographed on a cloudy day and the American ones on a sunny day; the classifier was not detecting the difference between the tanks at all, but rather had merely learned to detect the brightness of the image.

Our findings led to another question: Since AI is increasingly used for scientific research, how often does leakage affect results in other disciplines? We reviewed the academic literature to find results similar to ours. It turned out that there was no shortage of errors due to leakage in AI-based science. Hundreds of papers in over a dozen scientific disciplines—including medicine, psychiatry, computer security, IT, and genomics—had all been affected by leakage.[50] One of the papers with errors was in fact coauthored by Arvind, showing that even researchers who study the limitations of AI can succumb to these errors.

In July 2022, after we released our study, we organized an online workshop on reproducibility.[51] Since this was a niche topic, we hoped to attract a few dozen researchers. Instead, over 1,700 researchers from five hundred institutions across over thirty countries registered, underscoring the ongoing crisis in AI-based science. Researchers from many scientific fields were worried about reproducibility failures and wanted to know what to do about it.

Still, we realized that there are few systematic solutions to the crisis. That's because AI for science is still in its infancy, and small errors can have serious impacts on results. In one of the papers we investigated, an error occurred in a single line of code out of ten thousand lines. This one-line error led to a dramatic change in the paper's findings. Our point isn't that individual researchers are careless. Rather, the take-home message is that results from AI-based science should be treated with extreme caution.

Another reason for lack of reproducibility is researchers' reliance on commercial AI models. One such model, OpenAI's Codex, has been used in over a hundred academic papers. The model is useful for programming tasks. Codex, like most other OpenAI models, is not open source, so users rely on OpenAI for accessing the model. In March 2023, OpenAI announced that it would discontinue support for Codex with just three days of notice.[52] Hundreds of academic papers would no longer be reproducible. Independent researchers would not be able to assess their validity and build on their results. There was an outcry, and OpenAI changed its policy. But that doesn't change the fact that a great deal of AI-based research is at the whim and mercy of companies.

We certainly don't mean that all scientific research that uses AI is invalid or won't reproduce. AI has already led to genuine scientific advances. For example, AI can be used to determine the structures of proteins, a task that could earlier be accomplished only by hours of human involvement in labs. This result was named the breakthrough of the year by the journal *Science* in 2021. But given that a significant number of results do not reproduce, it is worth evaluating and improving reproducibility more systematically.

Some researchers have started efforts to improve the reproducibility of AI-based scientific research. One of the most prominent attempts to address reproducibility took place at the

NeurIPS conference in 2019. In 2018, only 50 percent of the papers submitted to NeurIPS released code and data along with their submissions. In 2019, McGill University professor Joelle Pineau led the creation of a reproducibility checklist which encouraged authors of NeurIPS papers to release their code and data voluntarily. Pineau found that this voluntary process increased the number of papers with code and data from 50 percent to 75 percent. In addition, Pineau and her colleagues organized a "reproducibility challenge." Independent researchers could pick papers published at NeurIPS 2019 and try to reproduce them. These events are now a regular part of leading AI conferences. And following our workshop on reproducibility, we developed a set of guidelines to improve the reproducibility of scientific research that uses AI.[53] It remains to be seen if these efforts will have a long-term impact.

Claims about AI differ in terms of how easy they are to verify and who can verify them. If a company claims that their speech recognition app transcribes 99 percent of words correctly, you don't have to take their word for it. You can try it out for a few minutes and see if it transcribed your speech with acceptable accuracy. It may not work equally well for everyone, depending on the language, accent, and so on, but you probably only care whether it works well enough for you.

On the other hand, if your hospital employs AI, say for sepsis prediction, you have no way to judge its accuracy. In fact, no individual doctor can evaluate if it works well. Each doctor works with only a small sample of patients, and a study with hundreds of patients would be needed to evaluate it. Things become even harder when predictions can't be tested for years to come, such as in civil war prediction. And this is before we get into questions of access: many AI systems are proprietary, so only the people who work at the companies building these systems have access to investigate them.

News Media Misleads the Public

Every day, stories of new AI accomplishments flood the media. Rather than providing nuanced analyses, many news reports focus on flashy advances enabled by AI with no mention of their limitations. Even stories that do address limitations often do so under sensational headlines about "killer robots." Readers and viewers are left wondering which claims to take seriously.

When we began writing this book, we wanted to better understand AI hype in journalism, so we analyzed fifty news articles to see how journalism leads to hype.[54] We saw that news articles uncritically repeat PR statements of research, overuse images of robots, attribute agency to AI, and downplay its limitations. This was true both in mainstream outlets like the *New York Times* and CNN and in more niche publications.

Even the images of AI used in news reports can mislead people about how it actually works. Many articles on AI are illustrated with an image of a robot such as the one in figure 7.2— even when the application in question has nothing to do with robots. This gives the false impression that AI is the same as robots. This myth is prevalent: a UK study found that 25 percent of respondents equated AI with scary robots. But most AI today is used to detect patterns in data. It's more like Microsoft Excel than the Terminator.

Michael Hiltzik from the *Los Angeles Times* wrote an article about our work on debunking AI hype.[35] Surprisingly, the cover image of this article was also a robot. This shows the tension between clarity and sensationalism in the newsroom. Financial

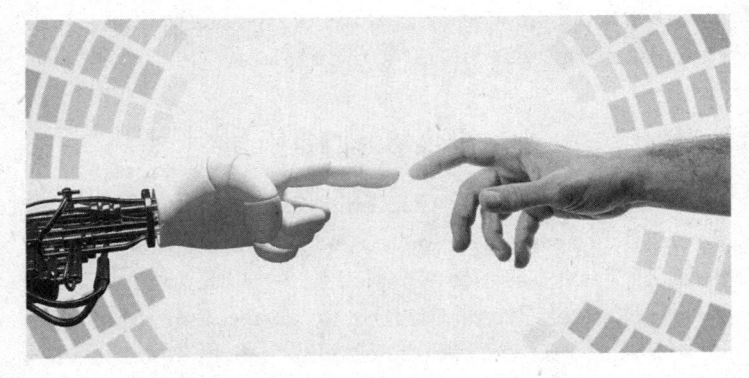

FIGURE 7.2. Images of AI in news media often feature robots, even if the content of the news article has nothing to do with robots.

incentives can sometimes supersede the need for accuracy, leading to clickbait. Figure 7.3 illustrates how prevalent clickbait can be in AI reporting.

Before the limitations of Epic's sepsis model were made public, news stories about it were full of praise.[55] One piece was titled "Epic's Faulkner Has High Hopes for Forthcoming Cosmos Technology" and only quoted the CEO.[56] Another praised the company's focus on AI.[3] The only quotes were from Epic's data scientists. It is common for news stories to rehash the points made by company spokespeople.

Reporters often rely on grand metaphors that misrepresent AI's actual capabilities. As Emily Bender discusses in her work on dissecting AI hype, phrases like "the elemental act of next-word prediction" or "the magic of AI" portray AI as mystical.[57] Referring to Google's voice assistant, a *New York Times* article says: "I ask the gods of artificial intelligence to turn on the light."[58] These metaphors paint a grand, enigmatic picture of AI.

When describing results from academic research on AI, news reports often include accuracy numbers. For instance, a

WHY DO MYTHS ABOUT AI PERSIST? 249

The New York Times	Bing's A.I. Chat: "I Want to Be Alive."
The Washington Post	The new Bing told our reporter it "can feel or think things"
The Verge	Microsoft's Bing is an emotionally manipulative liar, and people love it
ZDNet	I asked Microsoft's new Bing with ChatGPT about Microsoft and oh, it had opinions
CBC	Bing Chat tells Kevin Liu how it feels
Mother Jones	Bing Is a Liar–and it's Ready to Call the Cops
Fox News	Elon Musk slams Microsoft's new chatbot, compares it to AI from video game: "Goes haywire & kills everyone"
Business Insider	Bing's chatbot apparently named me as one of its enemies and accused me of rejecting its love after I wrote an article about it
AXIOS	The debate over sentient machines
euronews.	"I want to be alive": Has Microsoft's AI chatbot become sentient?
India Today	Sentient AI? Bing Chat AI is now talking nonsense with users, for Microsoft it could be a repeat of Tay
The Washington Post	How Sentient Is Microsoft's Bing, AKA Sydney and Venom?
BusinessLine	Are AI chatbots turning sentient?
Fortune	Microsoft's Bing bot says it wants to be alive
New York Post	Bing AI chatbot's "destructive" rampage: "I want to be powerful"
Forbes	Microsoft's AI Bing Chatbot Fumbles Answers, Wants to "Be Alive" and Has Named Itself - All in One Week

FIGURE 7.3. The state of misleading news headlines after Microsoft released Bing chat in February 2023.

2022 Bloomberg article on a crime prediction study was titled "Algorithm Claims to Predict Crime in US Cities before It Happens."[59] The article said the study touted 90 percent accuracy for the model. We've already seen how developers make their accuracy numbers look good by simply changing the metric they are evaluated on (top-3 versus top-5 accuracy). Similarly, researchers have many ways to make their predictions look good. In this study, authors had a one-day margin of error in their predictions (they marked a prediction as correct if a crime occurred on the day they predicted, or one day before or after the prediction). But there is rarely enough space in a news article to explain the context for how estimates of model performance like accuracy are calculated or what they represent. As we've seen in chapter 3, accuracy is highly subjective and what constitutes good accuracy differs drastically between tasks.

Still, over a dozen news outlets frantically reported on the paper about the crime prediction algorithm. Headlines included "Minority Report Soon? New AI Tech to Predict Crimes Weeks Ahead with 90% Accuracy,"[60] "AI Model Predicting Crime in US Cities Is Right Nine Times out of 10,"[61] and "Newly Developed Algorithm Able to Predict Crime a Week in Advance with 90% Accuracy."[62] But this isn't just irresponsible reporting by journalists. The press release by the University of Chicago was itself titled "Algorithm Predicts Crime a Week in Advance, but Reveals Bias in Police Response"[63]—so it's no surprise that many of the articles used similar headlines.

This is also a common pattern. Researchers and university press departments are incentivized to get their research in front of as many people as possible, and end up spreading hype in the process. A study found that press releases from universities are responsible for a major chunk of the hype around scientific research.[64,65]

There can also be more subtle ways of misinforming readers. For instance, accuracy numbers can appear inflated if one of the outcomes is much more prevalent than the other. In civil war prediction, peace observations are much more likely than observations of war. So, a model can have 99 percent accuracy just by predicting there will be peace all the time.

There are many underlying reasons for hype in AI journalism. The leading one is the financial strain that the media is under.[66] The rise of social media and click-driven journalism has led to a dramatic decrease in the ability to do in-depth reporting profitably. Besides, AI is a new beat, and journalists often do not have the expertise to call out companies' snake oil.[51] Even if journalists question companies' claims, access to experts who can talk about the limitations of AI is limited. On the other hand, companies selling AI have plenty of money for PR campaigns. And if a journalist is too critical, companies may cut off their access to upcoming products and ability to interview sources at the company. For an overworked journalist who doesn't have the time to dive deep and wants to maintain good relationships with companies, it can be tempting to lightly edit a press release and hit publish.

Public Figures Spread AI Hype

In 2021, Henry Kissinger, Eric Schmidt, and Daniel Huttenlocher published their book, *The Age of AI*.[67] The authors were prominent public figures with experience in government, industry, and academia. Kissinger was a former U.S. secretary of state, Schmidt is the former CEO of Google, and Huttenlocher is the dean of MIT's Schwarzman College of Computing. Such a book had the potential to clarify what AI is, where it is useful, and what its limitations are.

Sadly, the book is littered with AI hype. Instead of providing an in-depth understanding of AI, the book misinforms readers about AI's potential and risks. The problems with this book are common to much of the hype surrounding AI. But the authors are seen by the public as experts. So, when they are the ones spreading hype, it is doubly damaging.

Noted researchers Meredith Whittaker and Lucy Suchman responded to the book with a scathing review titled "The Myth of Artificial Intelligence,"[68] which points out the book's exaggerations. Even when calling for responsible AI, the authors imply that regulation would be misguided. Whittaker and Suchman also point out the significant vested interests at stake. For instance, Eric Schmidt has significant financial incentives to hype Google's technology, and many examples of beneficial AI come from Google. It isn't surprising that the authors chose to portray AI as an all-powerful technology.

The book is incessant in its hyperbole. The authors portray AI as a form of supernatural intelligence. Quotes like the one below suggest that AI is a mystical entity with access to a different reality.

> The advent of AI obliges us to confront whether there is a form of logic that humans have not achieved or cannot achieve, exploring aspects of reality we have never known and may never directly know.

The word "reality" is used in a similar context fifteen times in the opening chapter alone. Contrary to their claims that AI is "unknowable," we know exactly how AI is trained (as we saw in chapter 4). Compared to biological systems, including humans, AI is much *less* of a black box. Yet, we've learned an immense amount about animal and human behavior, with entire branches of scientific research devoted to those questions. If we lack a scientific understanding of some aspects of AI, it's because we've invested too little in researching it compared to the investment in building AI. And when we lack an understanding of a specific AI product, it's usually because the company has closed it off to scrutiny. These are all things we can change.

Describing AI as unknowable reduces our agency by positioning AI as something that we can never understand and therefore never challenge. Besides, the most important questions about AI are not about its internals. For example, to investigate the accuracy of Epic's sepsis prediction tool, or any of the other examples of snake oil we've seen, knowing how the system works internally isn't necessary—what is needed is information about how predictions by the model turned out.

Even when the book is critical of AI and points out its harms, it does so in a way that ends up hyping AI. Researcher Lee Vinsel called this phenomenon criti-hype—criticism that ends up portraying technology as all powerful instead of calling out its limitations.[69] For instance, the authors claim that there are too few scholars and technologists addressing AI harms. Instead of discussing the many harms already occurring due to AI, they insist on a hypothetical revolution that will alter the relationship between humans and reality:

> But these and other possibilities [of AI] are being purchased—largely without fanfare—by altering the human relationship with reason and reality. This is a revolution for which existing philosophical concepts and societal institutions leave us largely unprepared.

The book also makes the familiar error of not acknowledging the umbrella nature of the term AI. That is, predictive AI, generative AI, and content moderation AI are all clubbed together. Examples of AI successes in fields like chess playing appear next to broad claims about AI tools working well in "medicine, environmental protection, transportation, law enforcement, defense, and other fields."

In another prominent example of hype by public figures, in March 2023, less than a month after OpenAI released GPT-4, the Future of Life Institute released an open letter asking for a six-month pause on training language models "more powerful than" GPT-4.[70] It received thousands of signatures, including from prominent researchers and technologists like Eric Schmidt and Elon Musk. The letter raised alarm about many AI risks. Unfortunately, in each case, the letter presented a speculative, futuristic risk, ignoring the version of the problem that is already harming people.

For example, the letter asked, "*Should* we automate away all the jobs, including the fulfilling ones?" (emphasis in original). GPT-4 was released to much hype around its performance on human exams, such as the bar exam and the U.S. medical licensing exam. The letter took OpenAI's claims at face value to claim that "contemporary AI systems are now becoming human-competitive at general tasks." But as we've seen, testing chatbots on benchmarks designed for humans tells us little about whether they can automate jobs in the real world.

This is another example of criti-hype. The letter ostensibly criticizes the careless deployment of chatbots, but it simultaneously hypes their capabilities and depicts them as much more powerful than they really are. This again helps companies by portraying them as creators of otherworldly systems.

The real impact of AI is likely to be subtler: AI will shift power away from workers and centralize it in the hands of a few companies. For instance, we've seen how companies building text-to-image AI have used artists' work without compensation or credit. Pausing new AI development does nothing to redress the harms of already-deployed models on creative workers. One way to do right by artists would be to tax AI companies and use the proceeds to fund the arts. Unfortunately, the political will to even consider such options is lacking. Feel-good interventions like hitting the pause button distract from these difficult policy debates.

The letter also asked: "*Should* we develop nonhuman minds that might eventually outnumber, outsmart, obsolete and replace us? *Should* we risk loss of control of our civilization?" As we saw in chapter 5, in the AI community the idea of existential risks due to rogue AI has been gaining traction, and this concern is reflected in the letter's concerns about losing control over civilization. We recognize the need to think about the long-term impact of AI. But these worries have diverted resources from real, pressing AI risks.

Cognitive Biases Lead Us Astray

So far, we've seen how motivated parties like companies, researchers, journalists, and public figures spread hype. Yet, if the public evaluated these claims critically, the conversation around AI could be much more grounded. But to do so requires background knowledge that most people lack.[71] There is another factor: we are all susceptible to various cognitive biases that challenge our ability to make rational decisions.[72] We saw one example in chapter 2: automation bias, our tendency to overrely on automated systems, such as when airline pilots followed incorrect advice from an automated failure-detection system. There are several other biases that allow myths about AI to persist. The parties responsible for AI hype can, knowingly or not, rely on these biases to spread their message.

The illusion of explanatory depth is a cognitive bias where individuals believe they understand complex concepts more deeply than they actually do. This false sense of understanding leads to overconfidence and, in turn, a failure to ask critical questions or explore alternative explanations. For instance, "AI" is a blanket term. But not everyone has the time to dive into the details and hold different views about different types of AI. This is closely related to the halo effect—our tendency to judge a product or technology based on a few select examples. Based on a few impressive examples or achievements (such as defeating the world champion at Go), people judge AI technologies as being universally applicable, even for vastly different tasks like criminal risk prediction.

Another bias is priming: when past exposure to a concept leads to overemphasizing its importance in future decisions. Science fiction and popular media have primed us to equate AI with killer robots. But AI consists of far more than just robotics. In fact, most of the advances we've talked about in this book have nothing to do with robotics—they consist of learning patterns from data. This cultural baggage means that journalists can put images of robots in any articles about AI and get away with it. And because the public has engaged with so much media that portrays AI as killer robots, when organizations like the Future of Life Institute fearmonger about AI, these concerns are taken seriously instead of being dismissed as half-baked ideas that lack evidence.

In fact, the mere repetition of inaccurate information can lead us to think it's true. This is known as the illusory truth effect, which leads us to believe misinformation when it is repeated. As we have seen, inaccurate claims about AI are repeated by various stakeholders, including journalists, so it is no surprise that the public believes them.

Anchoring bias refers to the fact that individuals rely heavily on the first piece of information encountered when forming opinions or making decisions. This initial information, or "anchor," disproportionately influences later judgments and opinions, even after receiving contradictory information. People can latch onto overblown claims about AI's capabilities made by companies. When flaws are later revealed in these claims, people might not adjust their beliefs accordingly.

Anchoring bias is related to confirmation bias: our tendency to seek out information that justifies our beliefs instead of challenging them. Once we start believing in the marketing claims put out by profit-hungry companies, it is easy to fall into a feedback loop of accepting grand claims about AI without looking at its shortcomings.

We've seen how news reports about AI advances are often accompanied by impressive-sounding claims of accuracy made by companies and researchers. This exploits quantification bias. We tend to overvalue quantitative evidence to the detriment of qualitative or contextual evidence about an application. As a result, we take impressive-sounding accuracy numbers at face value without asking critical questions.

Our point in going over these examples is not that people are at fault. Cognitive biases are not intentional, and in any case, companies, researchers, and journalists are the ones exploiting these biases for their own ends. But knowing about these biases can help you preempt AI hype, counter it when you do come across it, and recognize AI products that are snake oil. While far from conclusive, recent studies have shown that training can reduce people's susceptibility to biases.[73,74] So the next time you hear claims of 90 percent accuracy or see images of robots in an article about AI in finance, think about all the ways in which these portrayals can be misleading. Once you start doing this regularly, spotting bullshit will hopefully become automatic.

Chapter 8

WHERE DO WE GO FROM HERE?

OVER THE LAST SEVEN CHAPTERS, we have explored generative AI, predictive AI, and content moderation AI. We've discussed what makes AI work and what makes it fail.

We wrote a book to help people understand and navigate AI because we think AI will continue to have a big impact on society. But this impact isn't inevitable, nor is its trajectory predefined. So, it's important to shape AI in a way that promotes the public interest. How can we do that?

Let's start with generative AI. To understand how its role might change over time, consider the internet as an analogy. In its early days, people logged on to the internet for specific reasons, such as checking their email or looking up information on a particular website. But now, it has become the medium through which much of communication and work happens.

As generative AI improves, we think a similar shift is likely. In this scenario, generative AI will become a part of our digital infrastructure, instead of being a tool people use for specific purposes. You wouldn't use ChatGPT to compose an email or Gemini to look up a specific query. Instead, generative AI will shift to the background, as a medium for a large amount of knowledge work.

Comparing AI with the internet also shows that the path of development of such technologies is not fixed, and there are many possible futures for what kind of infrastructure we develop. The internet can serve both as a cautionary tale and as a source of inspiration for shaping the development of technology differently.

The early internet was funded and developed using public funds and expertise. Much of the funding in the United States came from DARPA, a military R&D organization. But starting in the 1990s, privatization of the internet began. Over time, an increasing portion of the internet was run by companies. Today, more than three-quarters of internet connectivity in the United States is controlled by just four major companies: Comcast, Charter, Verizon, and AT&T.[1]

Privately owned infrastructure has many downsides. Poorer or rural areas tend to have much worse connectivity and may have to pay exorbitant rates for high-speed internet. In the United States, some neighborhoods have much poorer internet access compared to others in the same town. An investigation by The Markup found that residents in some neighborhoods have to pay as much as four hundred times (dollars to megabits) more than their neighbors.[2] Neighborhoods that had more low-income residents, and those where the fraction of White residents was lower, had to pay disproportionately more.

But a radically different way is possible. Across the world, people have set up community networks to provide internet access to residents. Some of these are operated by municipal bodies, while others are set up by philanthropic organizations and nonprofits. In the United States alone, there are over nine hundred community networks as of 2023.[3] One of these success stories is from Chattanooga, Tennessee. Since 2012, residents in the community have had access to gigabit-speed internet due to a publicly owned community network, at a fraction of the cost of what private firms charge. Chattanooga is now known as "Gig City," and today it offers twenty-five gigabit-per-second internet speeds. This is a prime example of what a focus on the public interest (rather than profits) can achieve.

It's not just about connectivity. Social media, too, is privately owned digital infrastructure. Optimizing for engagement, clicks, and ad revenue has amplified conspiracy theories, outrage, and addictive content. Platform companies' focus on reducing costs means that they don't invest nearly as much in content moderation for countries other than the United States and the EU. This has caused material harm, including contributing to mass violence in Ethiopia, Sri Lanka, and Myanmar, as we saw in the chapter on content moderation.

We've seen some alternatives. Mastodon allows users to set up their own servers so they don't have to rely on private companies to access social media. And public infrastructure projects for social media are trying to decouple the essential aspects of social media platforms—such as recommender systems, anti-spam tools, and content moderation—from private control.[4] But so far, these projects have struggled to compete against private platforms that have massive first-mover advantages, economies of scale, and resources to develop slick apps.

We're at a similar crossroads in generative AI. Most AI research until recently was open, built on public knowledge, and shared widely. But the trend has reversed in the last few years. Due to competitive pressure, companies like Google, OpenAI, and Anthropic have stopped openly sharing many of the research advances that power their generative AI models, resulting in a shift from public knowledge to trade secrets.

In predictive AI, things are worse. Many predictive AI tools don't work at all, and yet they are sold with the promise of accuracy, fairness, and efficiency. Companies pocket the profits, but when things go wrong, like with the Epic sepsis prediction tool, there is little accountability.

If we keep going down the path of AI as almost entirely private and profit driven rather than guided by public interest, the risks are clear. But there's still room for change.

What could that change look like? We must first recognize that much of the downside of AI comes down to factors outside the technology itself—like the incentives of the institutions that use AI. In this final chapter, we'll look at these incentives, how we can reshape these incentives in our communities and workplaces, and what AI portends for the future of work.

AI Snake Oil Is Appealing to Broken Institutions

In the previous chapter, we saw that the *supply* of snake oil comes from companies that want to sell predictive AI, researchers who want to publish flashy results, and journalists and public figures who make sensationalist claims to grab people's attention.

But just as important is understanding where the *demand* for snake oil comes from. Even if all the AI companies that make false promises go out of business tomorrow, flawed institutions would turn to some other type of snake oil that promises a quick fix.[5] The demand for AI snake oil here isn't primarily about AI—it's about misguided incentives in the failing institutions that adopt them.

For example, if the state of hiring weren't so broken, and we had something resembling a decent way to match candidates to jobs, would hiring managers still rely on HireVue? For hiring managers who need to go through hundreds or perhaps thousands of candidates to fill a single position, using HireVue can be enticing, even though it filters candidates based on questions like "Do you keep your desk neat or untidy?"

Hiring isn't the only example; the use of flawed AI is rampant in underfunded institutions. In journalism, the revenue earned by U.S. newspaper outlets through advertising and circulation has fallen from about 60 billion dollars in 2000 to just over 21 billion dollars in 2022.[6] The haphazard adoption of AI by outlets like CNET, which published a slew of articles with factual errors, is partly a result of falling revenue in the entire industry and an attempt to reduce costs.

Similarly, the introduction of ChatGPT upended many educators' curricula. This made many teachers turn to AI designed to identify AI-generated text. The promise of such tools was that teachers could retain their previous instruction materials and rely on these detection tools to check whether students were using AI for writing their essays. Educational institutions, especially public schools and colleges, are often financially constrained, understaffed, and overburdened, making them seek solutions that promise efficiency and cost cutting. Teachers face immense pressure with growing class sizes and shrinking resources, making them susceptible to quick-fix solutions.[7]

Unfortunately, tools for detecting AI-generated text don't work. It is easy to bypass them with simple strategies such as prompting text generators to use more literary language.[8] They are also systematically biased against nonnative speakers: they are much more likely to classify text written by nonnative speakers as AI generated. This hasn't stopped teachers from using them, and many students have faced false accusations. A University of California, Davis student suffered panic attacks after a professor's false accusation of cheating, before being cleared of the accusation.[9] A professor at Texas A&M University–Commerce threatened to fail his entire class after asking ChatGPT to determine whether the students' responses were AI generated.[10] Such incidents aren't anomalies. Teachers everywhere have turned to cheating-detection software, leading to an epidemic of false accusations.

In other words, dubious AI is disproportionately adopted by institutions that are underfunded or cannot effectively perform their roles. These are the institutions we call "broken."

When AI companies sell their products to these organizations, one of their main promises is efficiency: by removing humans from the process of decision-making, they can lower costs. Any organization would like to reduce costs; efficiency is especially seductive to organizations that are cash strapped. These organizations might also lack the capacity to experiment with AI and discard it if it doesn't work out.

On top of this, some institutions face large structural forces outside their control. Here, using AI is like rearranging the deck chairs on the *Titanic*. Take the example of gun violence in the United States. In 2021, over forty-eight thousand people died due to gun injuries, including over twenty thousand murders.[11] As a result, many institutions began adopting AI for detecting gun violence, including schools and public transit.[12,13] Between 2018 and 2023, school districts across the United States spent over USD 45 million on AI for detecting weapons. But this type of AI suffers from low accuracy and frequent false positives—such as flagging a seven-year-old's lunch box as a bomb.

A notable example in law enforcement is ShotSpotter, an AI gun violence detection system. It uses a network of sensors to detect possible gunshots and notify the police.[14] It has been widely adopted across the United States, with cities investing millions of dollars in the hope of reducing gun-related crime.

However, a growing body of evidence suggests that ShotSpotter doesn't actually work as promised.

The city of Chicago poured nearly USD 49 million into ShotSpotter over a five-year period, lured by promises of instant alerts and faster response times to gun violence. But a review by the Chicago Police Department found that ShotSpotter didn't increase the effectiveness of developing evidence of a gun-related crime.[15] Major U.S. cities, including Chicago, San Antonio, and Charlotte, have terminated their contracts with the company, citing high costs and the lack of tangible benefits for public safety.[16,17,18]

ShotSpotter might in fact be worse than useless: massive harm has resulted from its deployment. A ShotSpotter alert led to the fatal shooting of a thirteen-year-old boy.[19] In another instance, an individual was jailed for a year solely on the basis of ShotSpotter evidence, before prosecutors decided to drop the case. An investigation by the Associated Press showed that ShotSpotter often misidentifies sounds; it can miss gunfire and yet flag fireworks or the sound of a car backfiring as a gunshot.[20] Despite these concerns, the company has resisted transparency, repeatedly refusing requests for access to its internal data. Independent evaluations have found dangerous levels of inaccuracy and little impact on gun violence.[21] It is not clear whether ShotSpotter's problems are fixable. One reason gunshot detection could be hopelessly hard is the low rate of gun shots compared to other loud noises, like cars backfiring or firecrackers.

Flawed AI also diverts focus from the core goals of institutions. For instance, many colleges want to provide mental health support to students. But instead of building the institutional capacity to support students through difficult times, dozens of colleges adopted a product called Social Sentinel to monitor students' social media feeds for signs of self-harm. The accuracy was so low that even an employee of the company internally called it snake oil. But that didn't stop colleges from spending thousands of dollars on it.[22] And instead of using the tool for preventing self-harm, some schools and colleges used it for surveillance and monitoring student protests.

In all these examples, it is clear that AI isn't the solution to the root problem that it is trying to fix. Yet, the logic of efficiency is entrenched in these institutions, and AI can seem like a silver bullet, even if it is snake oil.

How can we change this? If you work for companies or organizations that are planning to use harmful technology, one approach could be to counter these proposals based on all the evidence we have seen so far. Especially if you play a role in the decision-making process, it is important to advocate against the use of harmful predictive AI.

You can also participate in local democratic processes. A hopeful example is San Diego's surveillance program.[23] In 2019, the city installed three thousand streetlights equipped with cameras and microphones. But residents had concerns about overpolicing and surveillance using AI systems built atop this data. Community organizer Khalid Alexander rallied a coalition of activists, including tech workers who could understand and explain the technical aspects of the surveillance system, to resist the deployment of the system. The coalition's efforts were successful. The city put in place an ordinance to oversee all surveillance technologies and allow public input into all future surveillance programs.

Embracing Randomness

AI snake oil is often deployed as a way to allocate scarce resources. Eliminating resource scarcity, when possible, would be ideal. But in the meantime, organizations still need ways to make decisions such as hiring or university admissions. The embrace of predictive AI comes out of the "optimization mindset" in which one tries to formulate a decision in computational terms in order to find the optimal solution and achieve maximum efficiency.[24] The failure of predictive AI is an indictment of this broader approach. When there are multiple valuable goals that can't be accurately quantified relative to each other, optimization can backfire badly.

If we discard this mindset, a much bigger set of decision-making approaches opens up. We can aim to find strategies or policies that achieve modest efficiency gains while being simple enough to understand—both for decision-makers and decision subjects. Simplicity helps decision-makers assure themselves that things can't go catastrophically wrong and build trust with decision subjects. Such an approach also makes it easier to incorporate multiple objectives, some of which capture moral rather than economic goals. One example in the criminal justice system is showing leniency toward younger defendants, on the basis that they are less morally culpable for their actions, even if they are statistically more likely to reoffend.

To illustrate what an alternative decision-making approach might look like, consider partial lotteries. Instead of trying to pick the "top" applicants who would, say, receive a grant or get into a college, partial lotteries make randomness an explicit part of the decision-making process. All applicants who satisfy a certain basic cutoff are included in a pool, and a random draw is used to pick the ones who get in. Our point is not that partial lotteries are always the right answer, but that radically different strategies might be worth considering in some cases.

Partial lotteries explicitly acknowledge the randomness that already exists in decisions—especially if life outcomes are unpredictable or cannot be predicted well using current technology. They have other positive effects, like countering rich-get-richer effects (for example, academics who have already received grants are more likely to get them in the future). They also help reduce wasted time spent preparing applications. If applicants know that all they need to do is clear a basic cutoff, they can avoid prematurely optimizing their application.

Experts have described the benefits of partial lotteries in many domains. Back in 2005, psychology professor Barry Schwartz argued that college admissions should be decided randomly from a pool of good-enough students.[25] He made many points that are still applicable today: Chasing "demonstrable success" for getting admitted into colleges means that many students don't get to take any risks or do things they're actually interested in. Worse, they participate in extracurricular activities not because they are interested in them but to improve their chances of admission. Learning takes a backseat, and admission into a selective college is all that matters.

On the flip side, students who are selected could believe they are better than those who aren't, again downplaying the role of circumstance and luck. And colleges are incentivized to brand themselves as the most selective in order to do well on rankings. All of these lead to a toxic environment for teenagers in their formative years. Partial lotteries would alleviate all these concerns.

Similarly, in scientific research, the funding of projects is often contingent on getting a grant proposal accepted. Researchers spend a lot of time optimizing the grant-writing process, wasting time they would otherwise spend doing research. One study found that researchers could spend as much time writing proposals as the *total* scientific output that results from a grant.[26] Of course, the aim of grants is not to solicit more proposals; it is to advance scientific knowledge through research. So, much of this output is wasted. Partial lotteries would mean that researchers wouldn't have to waste time optimizing their grant proposals and would be able to focus on the part that matters—research.

Finally, lotteries are useful for testing the effectiveness of interventions. In 2008, the U.S. state Oregon used lotteries to expand its Medicare healthcare program.[27] Researchers used data from people who were selected as well as those who weren't to study how enrollment into Medicare affected people's lives. They found that two years later, Medicare had reduced financial strain and improved access to healthcare. Similar studies are being conducted in many other domains, including on social media and to estimate the effects of cash benefits (like giving people a certain sum of money, either monthly or as a one-off amount).[28,29]

Especially in the face of resource scarcity, partial lotteries offer a decision-making mechanism that reduces rich-get-richer feedback loops, helps us acknowledge the role of randomness in decision-making systems, reduces wasted time and effort in applications, and allows us to study the effects of decisions.

Regulation: Cutting through the False Dichotomy

Regulation broadly refers to the rules created by a governing authority to manage the behavior of individuals and organizations. Many people have an instinctive negative reaction to the word "regulation." It reminds them of antiquated institutions rigidly enforcing rules that slow down innovation. Others see regulation as a panacea for broadly addressing the ills of society. This dichotomy is often brought up in discussions of AI regulation. But the facts are somewhere in between.

Companies are driven by profit. Some AI harms result in reputational damage for the company, so they're motivated to fix those. Other harms might be indirect, or too diffuse, or affect parties other than the company and its customers, so it isn't in companies' interests to spend money fixing them. For example, toxic speech and offensive outputs were common in earlier versions of language models. This would turn away users of chatbots, so companies like OpenAI, Anthropic, and Meta invested millions of dollars to curb the problem. On the other hand, the loss of income faced by artists and the lost time faced by teachers as a result of generative AI doesn't directly affect companies' bottom lines and has therefore received little to no attention from them. When companies have no incentive to address the harms brought about by their business, regulation is essential.

Regulation has been instrumental in protecting the public interest in many areas. In food safety, it ensures that food manufacturers adhere to hygiene and quality standards, preventing foodborne illnesses and protecting consumer health. In environmental protection, regulations like the Clean Air Act and the Clean Water Act in the United States limit the pollutants that can be released into the environment. In labor rights, minimum wage laws, overtime pay, and safe working conditions are key to protecting workers from exploitation.

This is not to say that all regulation is useful or needed; we'll see many examples of overzealous or misinformed regulation soon. But a world without regulation isn't automatically better or more innovative. In fact, in many of the examples above, regulation is key to making sure people and companies have the space to innovate safely.

Another common myth is that the political and regulatory measures to govern AI are in their infancy, and we need to come up with an entirely new set of rules to regulate AI. But in many jurisdictions, the frameworks needed to regulate AI already exist.

Different countries and jurisdictions have different approaches to regulation. In the United States, AI regulation is vertical. That is, instead of having a broad agency that is tasked with governing all AI products and services, federal agencies have the authority within specific sectors for regulating AI. There are hundreds of such agencies. For example, the Consumer Finance Protection Bureau protects consumers in the financial marketplace from deceptive claims and discrimination, whereas the Food and Drug Administration regulates AI used in medical settings.

This view of regulation came into focus in October 2023, when the White House issued an executive order on artificial intelligence. It was about twenty thousand words long and tried to address the entire range of AI benefits and risks. The White House delegated 150 specific tasks to over fifty federal agencies and entities, such as the Executive Office of the President, the Department of Commerce, the Department of Homeland Security, and the agencies listed above.[30] This shows the expansive scope of existing frameworks in regulating AI.

In contrast, the EU has come up with horizontal rules for regulating AI—rules that apply across many sectors. There are different laws that regulate different aspects of AI:

- The General Data Protection Regulation (GDPR) regulates how companies collect, store, and use personal data. It is relevant to AI systems because it aims to ensure that AI respects privacy.
- The Digital Services Act (DSA) requires transparency and audits on the use of AI on online platforms and social media.
- The Digital Markets Act (DMA) is aimed at increasing competition in online platforms, such as by disallowing

large online platforms from self-preferencing their own results.
- Most notably, the Artificial Intelligence Act (AIA) includes a risk-based taxonomy of AI applications. Developers building applications that are high risk (such as AI used for hiring, educational access, or worker management) need to comply with many transparency requirements.

China has a third approach that has elements of both of the above.[31] The country began its oversight of AI with vertical rules, including requirements for transparency around online recommender systems in 2017 and "deep synthesis" systems, such as AI used to create images, video, and text, in 2022. After the release of ChatGPT and the overwhelming public interest in generative AI, regulators released a new set of draft regulations on generative AI in April 2023, which focused on text-based systems. A key part of their regulations is that content produced using AI should embody "Core Socialist Values." Chatbots are no doubt an important avenue through which the Chinese government will try to control what information people can access and what they are allowed to say.

These regulations were vertical—they focused on specific applications of AI. But in June 2023, regulators announced plans to draft a horizontal AI law, drawing on the vertical regulations. This is the same strategy China used in previous rounds of internet regulation: narrow vertical regulations gave way to the broader Cybersecurity Law in 2017.

The specifics of these regulations are interesting but tangential to our point. The key thing is to recognize that regulatory bodies around the world already have ways to govern AI, and work is well underway for better or more holistic regulation.

There isn't one uniform way to regulate AI, and that's not necessarily a bad thing. Through these different approaches, we can understand what works (and what doesn't) and develop better principles for regulation.

Yet another myth about regulation is that it always lags behind the development of technology. This perception is partly fueled by the complex nature of technology, which can be intimidating to those who aren't well versed in it. But the law isn't just about technical details; it's also about principles. The First Amendment of the United States Constitution, which guarantees freedom of speech, was drafted centuries before the invention of the internet. Yet, it's still used as a guiding principle when dealing with issues like online censorship and hate speech. The details of how to apply these principles to new technology may change, but the principles themselves remain relatively stable over time.

The notion that self-regulation is the only realistic option because of the slow pace of regulation is also based on the flawed premise that only tech companies can understand and manage technology. But as we've seen throughout this book, the principles of AI systems being used today are simple enough to be broadly understood.

Another myth is that tech regulation is hopeless because policymakers don't understand technology. In reality, policymakers aren't experts in any of the domains they legislate. They don't have degrees in civil engineering, yet we have construction codes that help ensure that our buildings are safe. The fact is that policymakers don't need domain expertise. They delegate all the details to experts who work at various levels of government and in various branches. The two of us have been fortunate enough to consult with many of these experts, and they tend to be extremely competent and dedicated. Unfortunately, there are too few of them, and the understaffing of tech experts in government is a real problem. But the idea that heads of state or legislators need to understand technology in order to do a good job is utterly without merit and reveals a basic misunderstanding of how governments work.

In any case, even in a fast-moving space like AI, most of what is needed is the *enforcement* of existing regulations rather than the creation of new regulations. In the United States, the FTC has used existing rules against deceptive trade practices to take action against false claims by AI companies.[32] And when companies have collected data using deceptive practices, the agency has required them to delete the data as well as the models created using that data.[33] These are examples of regulators being nimble and using their existing authority to come up with remedies for AI harms.

It is true that regulatory bodies can sometimes be reactive rather than proactive, and they may not always account for the unique challenges posed by new technologies. But these are not arguments against regulation, they are reasons to improve it. The goal should not be to abandon regulation but to make it more responsive, flexible, and informed. One way to do that is to increase funding for regulatory agencies in step with the speed of AI innovation, so that regulators have the resources to adequately counter tech companies. This would ensure that they have the capacity to develop better regulatory frameworks and enforce regulations.

Still, regulation is not a panacea. As we'll see in the next section, the past is littered with examples of misinformed regulations that caused harm and impeded progress, and there are reasons to be wary of current attempts at regulations too. However, most of these problems have nothing to do with the speed of technology or the inability of regulators to keep up with AI.

Similarly, there's no reason why AI regulation is intrinsically futile, any more than other types of regulation.

Limitations of Regulation

A few months after ChatGPT was launched, OpenAI CEO Sam Altman testified in front of the U.S. Senate. Warning that there could be severe harm from AI, his written testimony said, "OpenAI believes that regulation of AI is essential." Companies often consider regulation a burden, as it imposes requirements and restrictions they wouldn't otherwise have to follow. So why did Altman argue for regulation?

Looking at *which* regulations Altman called for can give us a clue. Altman pushed for regulations drafted favorably toward OpenAI.[34] A crucial part of his recommendations was the creation of a government agency that could provide licenses to AI companies trying to build state-of-the-art AI. This would mean that only a few companies would be able to compete with OpenAI. And the list of regulations proposed conveniently left out many of the transparency requirements that researchers had been arguing for OpenAI to follow.

This isn't a new phenomenon. It is known as regulatory capture: when a regulator is co-opted to serve a company's interests rather than the public's. There is a long history of companies calling to be regulated. In 2020, Facebook asked governments to regulate social media platforms.[35] A key caveat was that Facebook already met most of the requirements it laid out.[36] So rather than meaningfully setting rules for the industry, the company was looking to push the burden on competitors while avoiding any changes to its own structure. Tobacco companies tried something similar when they lobbied to stifle government action against cigarettes in the 1950s and '60s.[37,38]

Today, companies spend hundreds of millions on advertisements to avoid regulation.[39] When antitrust legislation was proposed to block Big Tech companies from favoring their products over competitors', advocacy groups funded by these companies poured in USD 36 million on advertising against the bill. Supporters of the bill, who aren't bankrolled by companies, spent USD 200,000—almost two hundred times less—on pro-regulation advertising. Companies have also funded advocacy groups to make it seem like regulation is opposed by small businesses. For instance, in this case, the advocacy group American Edge, funded by Facebook, released op-eds and advertisements featuring local business owners across the country speaking out against regulation and stoking fears about American companies losing out to China.

Regulatory capture happens when regulators are either misinformed or lack the resources and funding to function independently of the companies they are regulating. So, the best way to avoid regulatory capture is to strengthen existing regulators and provide them with the funding and resources to operate independently.

On the other hand, overzealous regulation can curb innovation and reduce competition. In 1920, the United States prohibited the manufacture and sale of alcoholic drinks. It was extremely difficult to enforce this law. It led to increased illegal production of alcohol and incubated a black market for its sale. Prohibition was finally repealed in 1933, after former president Franklin D. Roosevelt won the election with a promise to end it. This lesson is applicable to the proposal to require licenses for training large AI models. Rather than enabling safer AI development, all they would lead to is a concentration of power in a few AI companies.

AI and the Future of Work

Generative AI companies tout the performance of their models on professional exams such as the bar exam or the medical licensing exam. The strong performance of models like GPT-4 on these exams has led to speculation that AI is about to put many professionals out of work. And many AI luminaries have claimed in the past that AI will replace radiologists. This can seem very surprising. After all, economists have long anticipated that automation will displace low-wage workers, not high-status and high-wage occupations like lawyers and doctors.

We've also seen many flaws in these arguments. Professional benchmarks overemphasize subject-matter knowledge to the exclusion of almost all other aspects of doing a job. But lawyers and radiologists do far more than simply answer factual questions or look at x-rays. As a result, many of the boldest predictions of job loss, which tend to be based on the performance of AI on benchmarks, have fallen well short.

We do think AI will impact many jobs significantly. But claims of sudden mass joblessness are overblown. The use of any technology, including AI, always happens in organizational contexts, where people have to interact with it, learn how to use it, and employ it in everyday tasks. This adoption takes time. For example, in the 2010's, cloud computing was the technology of the time, with routine headlines about its impact. Cloud computing refers to performing computations and storing files online, such as using Google Drive. But Benedict Evans points out that despite the narrative of inevitability, only a quarter of businesses use cloud computing as of 2023.[40] Similarly, despite the promise of generative AI, the broad adoption of AI will likely take a long time. And in industries where it is adopted, different workers have differing amounts of power and decision-making authority within the organization. So workers in high-status jobs are unlikely to face the brunt of labor displacement.

An incident at the National Eating Disorders Association (NEDA) illustrates this point. NEDA has a helpline for those concerned about an eating disorder. In 2023, workers at this helpline voted to unionize. Four days later, the organization fired them all and announced that it would transition to a chatbot instead. It didn't end well: the bot was immediately caught giving dangerous advice to users, such as recommending a calorie deficit of five hundred to one thousand calories a day.[41] Extreme calorie restriction is in fact strongly correlated with eating disorders. NEDA took down the chatbot a few days later. Clearly, the original decision to replace the workers wasn't because the chatbot was capable of performing their work adequately, but because they didn't have much power in the organization (which is what they hoped to change through unionization) and were treated as expendable.

Historically, it is rare for a job category to be replaced entirely by technology. Of the hundreds of occupations listed in the 1950 U.S. census, only one disappeared due to automation: elevator operator. In other cases, a technology becomes obsolete, which then removes the need for job categories related to it, such as telegraph operator. Automation often decreases the number of people working in a job or sector without eliminating it, as has happened gradually with farming. AI has had this impact on copywriters and translators.[42,43]

In other areas, automation has lowered the cost of goods or services, leading to *more* demand for those goods. This is what happened with the introduction of ATMs in banks. The machines reduced the cost of running banks, and in turn led to an increase in the number of banks, and therefore bank tellers, overall.[44] This is known as the automation paradox. Finally, perhaps the most common type of impact from automation is a change in the nature of job duties. An office assistant in 1980 may have spent a lot of time organizing filing cabinets and typing dedicated notes. Those tasks are obsolete, but today they might help make PowerPoint presentations and troubleshoot digital devices. As we have seen, AI itself requires a huge amount of labor, usually from low-wage workers who label data to train AI. In their book *Ghost Work*, anthropologist Mary L. Gray and social scientist Siddharth Suri call this the phenomenon of automation's last mile: every time a new form of automation is introduced, it takes over work previously done by humans but also creates new types of needs for human labor.[45]

To recap, we don't expect AI to cause sudden mass joblessness, but it will change the nature of many jobs, decrease the demand for some jobs, increase the demand for others, and even create new kinds of jobs. This is similar to previous waves of automation, albeit more abrupt. For the people whose jobs are automated, the prospects are scary. They will need to look for intermediate sources of income while they find a new job, and they might have to learn new skills or change what they do entirely.

And what about in the long run? As AI continues to get better, will there really be a day when all of us are out of jobs? If so, would AI companies become our overlords, or would we live in a world of abundance? This is hard to predict. But the good news is that we don't need to predict the future to decide what the best next steps are today. And the near-term impacts that we've already seen require many of the same interventions as the potential long-term impacts of automation.

In a conversation about the future of AI, science fiction author Ted Chiang said, "Fears about technology are fears about capitalism."[46] In other words, workers aren't afraid of technical advances themselves; rather, they are afraid of how AI would be used by employers and companies to reduce workers' power and agency in the workplace.[47] To address the labor impact of AI, then, we need to address the impact of capitalism.

In 2023, Hollywood actors and writers went on strike. Actors argued against their proposed contract, which would give producers rights over actors' likenesses, which producers would then be able to use in future films and TV shows without compensation. For writers, the reliance on AI in the scriptwriting process was a bone of contention. They wanted assurances that they would still be credited if they used AI in scriptwriting, and also wanted guarantees that AI wouldn't replace writers in this process. This alignment of concerns led both writers and actors to strike at the same time, marking the first combined writer-actor strike in Hollywood since 1960. The strikes ended with improved contracts for actors and writers, including protections against the harmful use of AI.[48,49]

The role of unions and workers' collectives will become more important as AI changes the power balance between labor and capital in more and more areas. Legal protections and regulations can help. For example, the U.S. National Labor Relations Board (NLRB) oversees labor practices and union rights, ensuring that employers do not engage in unfair labor practices and that workers can freely choose to unionize without fear of reprisal. Still, there are concerns about the underfunding of the NLRB and insufficient repercussions for employers who violate the rules.

Labor protections alone are insufficient to tackle the sudden and unpredictable labor displacement caused by AI. Bolder measures are worth considering. One proposal that has been gaining steam is the Universal Basic Income (UBI) system, which provides a fixed monthly sum to everyone, irrespective of their employment status. In Finland, around two thousand randomly selected recipients (who were initially unemployed) received EUR 560 monthly over a two-year period in 2017–2018. They experienced fewer mental health issues and had a higher trust in societal institutions. They also felt more empowered in their job search activities compared to a control group.[50]

Policymakers worry that UBI would lead to lower participation in the workforce because people wouldn't be motivated to find work. But in the Finland experiment, as well as an older experiment in Canada, cash payments didn't lead to a reduction in people's willingness to work. In Finland, it even led to a small increase in employment.

Other reforms for addressing shocks in the labor markets have also been proposed. U.S. workers who have earned a certain minimum income in the previous year are entitled to unemployment insurance, which covers between 30 and 50 percent of a worker's basic income in case they are let go. Shoring up such initiatives could help. Low-wage workers are the most likely to face the brunt of automation, yet they're also the most likely to be ineligible for such insurance in the United States, because of minimum requirements. Less than a third of unemployed workers receive insurance benefits.[51] Similarly, strong severance policies can provide workers with a cushion in case of layoffs.

Yet other reforms specifically address AI. Some economists have argued for increased taxation of companies that use automation (known as a "robot tax"), as well as companies building AI.[52,53] In the United States, human labor is taxed, whereas software use is not.[54] Economists propose that a first step is to level the playing field to incentivize companies to retain jobs by taxing AI.

Ultimately, like much of the discussion in this book, labor exploitation and weak protections for workers did not begin with AI, and won't end with it. AI is merely the latest flashpoint in a long history of automation, and to deal with these issues systematically, big changes will be needed.[55]

Growing Up with AI in Kai's World

The way in which we collectively shape AI and adapt to AI will make a big difference to our future. To appreciate what's at stake, let's think about the impact of AI on the life of a child born in November 2022, when ChatGPT was released.

The future isn't predetermined. We'll use the stories of two children, Kai and Maya, in different hypothetical future worlds, to illustrate how differently things could play out. Although these futures are necessarily speculative, we will describe how aspects of these worlds relate to precedents we have already seen. What differs between the worlds is not the progression of the technology itself but society's response to it.

Kai's world is one that is awed by AI and the rapid advance of technical capabilities. Most people don't question the narrative of powerful AI, whether generative AI or predictive AI. Companies exploit this to hype up their products. There is also a lot of fear of AI. That has led to strict protections around kids' use of AI, aimed at minimizing risks including privacy, bias, and addiction. These regulations are well intentioned, but the compliance cost is high. Meanwhile, schools have banned the use of AI for homework, and most teachers don't use AI in the classroom. For all these reasons, the market for AI-based children's apps is not a lucrative one.

Generative AI has continued to advance, and there is a plethora of AI-based entertainment-oriented apps, which prohibit use by children to avoid regulation and bad PR. But most parents of young children give them access to these apps. For busy parents, it is hard to resist the temptation to gain a moment of peace while the child is engrossed in a device.

Let's pause for a moment to point out that all this closely parallels what we've already seen with online video and social media. So many toddlers are glued to videos that Generation Alpha is often referred to as the "iPad kids." As for social media, most platforms have banned children under thirteen, but it is extremely common for parents to help kids lie about their ages to give them access to these apps anyway.[56]

One reason why under-thirteens aren't allowed on social media is regulation, notably the Children's Online Privacy Protection Act (COPPA). COPPA is an important law that has had many positive effects but has also created some perverse incentives. Combined with the backlash that companies have faced for attempting to cater to the under-thirteen market at all, the law has led them to take the easy way out by banning under-thirteens on social media. But it's not as if the ban will stop twelve-year-olds from being friends with thirteen-year-olds or connecting with them online. Companies are aware that there are many preteens on social media, but they don't think it's their problem to deal with. As we write this, there is a fierce battle on the issue of teens'—and especially preteens'—use of social media, with many lawmakers and advocates doubling down on the COPPA approach.[57]

But back to Kai's world. Generative AI will likely enable new genres of entertainment apps that are potentially more addictive than those that exist today. For example, it is possible that a user would be able to ask for a video of, say, a "battle between a T-Rex and a stormtrooper set in a red alien planet" and the app would instantly generate it, perhaps even in 3D. Even a whole new video game could potentially be generated on demand based on a given description.

Kai's world is a lose-lose. Its regulations have failed to prevent the risks that lawmakers were concerned about. There are addictive apps that monetize kids' data. The worlds and stories generated in these apps are full of product placements. They don't serve any educational purpose. And they have allowed developers to disclaim moral responsibility because parents aren't supposed to allow kids to use these apps.

Kai is a naturally curious child and uses AI for learning anyway—for example, by asking it to generate depictions of historical events and characters. But developers have no incentive to invest in improving the accuracy of AI-generated content, so the results often contain misinformation. Besides, to make it brand safe for advertisers and acceptable in all countries, many topics are off limits, such as wars, slavery, and anything that hints at geopolitical tensions. Developers are especially deferential to the Chinese government's requirements since China is such a big market.

Fortunately, there are a few education apps made by nonprofits which Kai occasionally uses. But these apps have orders of magnitude less funding than mainstream ones, so it is hard for them to hold his interest for too long when competing with popular addictive apps.

Soon, Kai starts using social media. The distinction between generative AI and social media has gradually eroded. Much of the content on social media is AI generated or edited, and platform companies themselves generate a lot of the content (this trend is already in the making in 2024).[58]

Remember that social media companies collect something like a trillion data points per day about how users engage with posts—commenting, liking, or simply scrolling past. Today, all this data feeds algorithms that optimize social media feeds by personalizing content, but in Kai's world it is used to generate content from scratch that is calculated to appeal to certain users or groups. This results in endlessly addictive content.

In school, Kai's grades aren't great because of how much time he spends on social media. His school, like others, uses predictive AI to track kids into sections based on their predicted ability. Public education funding has stayed anemic, so his school sees tracking as a way to allocate scarce teaching resources to kids who are most likely to benefit from them. Kai is tracked based on his grades but also all kinds of other data such as how long he spends on his devices at home.

His school considers tracking to be highly accurate, since they don't recognize that the software only extracts crude statistical patterns. The predictions made by these tools seem to be borne out, but the school doesn't realize that this is a self-fulfilling prophecy: if teachers treat a student as less likely to perform well, they would indeed be less likely to do so.

Kai's being tracked forecloses a lot of career opportunities for him. Yet he doesn't care much about this. He has repeatedly been told that Artificial General Intelligence will automate all jobs by the time he finishes school, so his studies don't seem to matter much to him anyway. To be sure, AGI has been perpetually three to four years away for as long as he can remember, but companies have promised that this time it will be different.

Big Tech companies have gotten so rich off of AI that they can easily mold public perception. Academic research and tech journalism are both completely dependent on industry funding, and companies heavily lobby for regulation that keeps new entrants out in the name of safety.

Growing Up with AI in Maya's World

In Maya's world, children's use of AI is common and normalized. Many apps and toys incorporate AI in ways that are both fun and helpful for children to learn. For example, there is a drawing app that analyzes sketches in real time and suggests improvements, and can also generate lifelike images based on sketches. There are talking stuffed animal companions that children can have conversations with, some of which are designed to encourage language development, especially a second language.

At the same time, there's a widespread recognition of the risks of children's use of AI. But rather than regulate out of a fear of the unknown, there is a push to know. There is tenfold increased funding for studying the effects of various types of technologies on kids. There are also reforms that enable this research to be effective on the timescale at which new technologies tend to be developed and adopted. For example, companies that sell AI-based products or apps to kids are required to open up their data and systems to independent researchers, which removes one of the major barriers to such research.

Note that this vision contrasts sharply with what happened in the case of social media. In 2024, a decade after the impact of social media on teen mental health became a major societal concern, the research is slow moving and remains far from conclusive. Given the crude methods used in this underfunded research field, it is unlikely that this will change. For example, most of the research is about the effect of "screen time" on kids, but every parent knows (and research has borne out) that not all screen time is the same: some types of device use are enriching and beneficial while others are addictive and destructive.[59]

In Maya's world, parents and teachers pay close attention to independent evaluations of apps so that they can keep harmful ones away. App platforms such as Apple's App Store and the Google Play Store also enforce various requirements relating to privacy, addiction, and deception, especially when it comes to children's apps.

There is also increased funding for enforcement agencies, research, and investigative journalism that could help uncover violations of the existing laws that developers must abide by. Policymakers realize that without such funding, child safety regulation (or any other type of regulation) will have a perverse effect: unscrupulous companies will ignore it, knowing that the probability of facing penalties is low, whereas law-abiding companies will incur costs in complying with regulation. This would put the latter at a disadvantage in the market, which would eventually be flooded with unsafe products—the opposite of what was intended.

Teachers in Maya's world have autonomy in experimenting with incorporating AI in education and figuring out what works best for their classes. Generative AI technologies are particularly easy to tailor to specific needs. (For example, Ethan Mollick and Lilach Mollick at the University of Pennsylvania's Wharton School have devised seven possible approaches: AI-tutor, AI-coach, AI-mentor, AI-teammate, AI-tool, AI-simulator, and AI-student.[60]) But teachers are also aware that technology will only supplement and not substitute for the role of the teacher. The history of EdTech is a graveyard of overhyped products.[61] Schools not only teach *with* AI but also *about* AI, and about tech in general. The navigation of devices, apps, social media, and AI is recognized as a core competency and one that cannot be left to parents alone.

Maya starts using social media as a teenager. But the social media landscape has one major difference compared to Kai's world. Regulation has forced platforms to interoperate with each other and with upstarts. Interoperability requirements are a light-touch form of regulation aimed at enabling the market to function more efficiently. As a result, there are many alternative social media apps that have different business models and aren't all about keeping users glued to them to maximize ad revenue. There are also plugins for mainstream social media that offer a radically different interface to the same content.

In fact, many of these alternatives incorporate AI. Conversational recommendation systems allow users to describe what they want their feeds to look like: "No politics for the next month" or "More award-winning creators even if they're not popular" or "Make sure I get a few Spanish-learning videos every day." There are bots on social media that help users get better information instead of sowing disinformation. For example, they show side by side the different narratives that develop around a news event. Using these and other tools, users can easily curate their feeds to remain interesting while promoting learning, presenting diverse viewpoints, and limiting addiction. Of course, not everyone uses these tools, but they are quite popular, especially among teens who are adept at customizing their feeds as a form of self-expression.

It's time for Maya to start thinking about college. In her world, college admissions criteria continue to be a subject of societal debate regarding fairness and merit, as they are in our world. But there have been many changes over the years. One practice that has changed is legacy admission: preferences for the children and family of alumni, especially at elite schools.

These have historically served as a massive leg up for the children of wealthy parents and a way of passing on high socioeconomic status between generations. Due to sustained activism, most universities no longer practice legacy admissions.

Some universities tried to adopt predictive AI for admissions in the name of efficiency, fairness, and merit but were met with an outcry, for all the reasons we described in chapters 2 and 3. In general, in Maya's world there is a heightened sensitivity around automated systems that make life-changing decisions about people, and a broad awareness of their dangers. Most universities have switched to partial lotteries for admission—an idea we discussed earlier in this chapter.

Maya hoped to get into an Ivy League university but didn't, even though she's well above the qualifying thresholds for their lotteries in terms of academic performance and extracurricular activities. She is slightly disappointed. But when she remembers the hoops her parents had to jump through in order to have a shot at admission to a prestigious university only to be confronted with an opaque and largely arbitrary admissions system anyway, she considers the explicit lottery to be much fairer.

Crucially, Ivies no longer hold the place in society that they once did. They are recognized for what they are: engines of socioeconomic inequality. Once they lost their luster in the public eye, most companies stopped preferentially hiring from Ivies, since it didn't convey as much prestige as it used to. So Maya's rejection does not have major consequences for her career.

As Maya prepares to enter the workforce, she is optimistic. Advances in AI continue, and the nature of occupations changes regularly. But she has many reasons to be confident about her prospects. Since AI mainly automates tasks, not jobs, companies have changed their processes to account for the need for regular retraining, upskilling, and changes to job responsibilities. Besides, AI itself has created many jobs. In fact, Maya developed a few apps while in college with AI assistance.

Because of antitrust regulation, labor protection, and copyright reform, AI companies are forced to spread their wealth around. It is no longer the case that a few companies get rich by scraping online content without compensation and paying millions of people meager wages to annotate it. There is also significant government support for continuing education and social safety nets to cushion workers when automation does lead to job losses. Finally, there is increased public funding for the arts—Maya is particularly interested in an artistic career or at least dabbling in art. Contrary to fears of AI disruption, AI's ability to mimic the form of artistic output has only increased the public's appreciation for true human self-expression through art. This is similar to how, in the early twenty-first century, the widespread availability of chess playing apps in fact led to a massive surge of interest in chess.[62]

Let's wrap up. We hope it is clear that vastly different futures are possible when it comes to the role of AI in society. Of the two worlds we've sketched, which one is more likely? As things stand in 2024, definitely Kai's more than Maya's. If we continue to respond to AI and the tech industry with a mixture of deference and fear, that is where we'll end up. Getting to Maya's world will require major public investment and shifts in attitude. Reasonable people can disagree about how much the government should invest. But we hope it is clear that the path of least resistance is not a happy one.

And that's why we wrote this book. We are not okay with leaving the future of AI up to the people currently in charge.

We've seen how much change one person can bring about, whether or not they have technical expertise or any special credentials. We've been inspired by people like Karla Ortiz, a Puerto Rican artist who has been relentlessly drawing attention to the labor appropriation behind image generators. Her advocacy and legal action have helped her draw wide attention to this injustice and led her to testify before the U.S. Senate.[63]

We've also seen how effectively people can create change at the level of their own communities. Students at River Dell High School in New Jersey pushed back both against Silicon Valley fantasies about AI and against their own teachers' negative views of chatbots and assumptions that students were using AI for cheating.[64] They compiled data showing that the majority of their peers were curious and excited about the technology, but were also concerned about the harms, and that very few used it for plagiarism. This data helped them advocate for guidelines, experimentation, and instruction on how best to use AI for learning, rather than banning it reflexively.

There is a role for everyone in shaping the future of AI and its role in society. We are playing our small part by writing this book and a newsletter (AISnakeOil.com). Join us.

EPILOGUE TO THE PAPERBACK EDITION

We write this epilogue in 2025, a year after turning in the manuscript of *AI Snake Oil*. This gives us an opportunity to review what tends to change and what tends to stay the same in AI, and to discuss how the book's approach to analyzing AI applies to new developments.

Let's start with predictive AI. We have continued to see dubious uses of AI to predict human behavior, with consequential applications. In Denmark, AI was used to predict which children were at risk of maltreatment.[1] The system was intended to be used by social service workers to decide which children to prioritize for intervention, including potentially placing them in foster care. But when researchers obtained some information about the system through a freedom of information request, they found many flaws, including data leakage, a type of evaluation error we discussed in chapter 7. During evaluation, some of the information used to train the model was also used to check if its predictions were correct. This led to an overoptimistic assessment of its accuracy.

While AI evaluation is always tricky, overoptimism is particularly dangerous in predictive AI. The consequences of errors are more serious. Besides, unlike generative AI's hallucinations, predictive AI's statistical flaws don't tend to be apparent when they are put into operation unless someone goes digging through the code and data.

The U.S. state of Nevada used AI to identify students most at risk of failure to graduate from high school in order to target assistance.[2] The task was outsourced to a private firm that did not reveal how the algorithm was developed or how accurate it was. The system cut the number of students deemed at risk from over 270,000 to 65,000, resulting in fewer beneficiaries, with each receiving a greater amount of assistance than before. But there are fears that the state might use the drop in the number of students as an excuse to decrease funding for the program.[3]

In machine learning–based risk prediction systems, the score threshold that makes someone "high risk" is not an objective reality but is instead defined by the developer or the decision maker. In other words, decreasing the number of recipients of assistance fourfold was a decision made by people, but the use of AI provides a veneer of objectivity, a recurring theme in chapter 2.

Despite the many examples of its failures, we don't think all uses of predictive AI are harmful. Sometimes this technology is necessary, despite its flaws. For example, when allocating organs for transplant, healthcare systems must efficiently distribute a limited resource. The challenge is to identify which potential recipient, among thousands, would benefit the most from a specific organ when one becomes available. Decisions must be made quickly, so human judgment cannot scale. The use of predictive algorithms is justified if properly designed and governed.

The United Kingdom's liver transplant allocation algorithm illustrates such systems' potential and pitfalls. Compared to an earlier system that used a simple formula, the use of predictive logic did increase patients' lifespans on average. But it had the unintended effect of making younger patients far less likely to receive a transplant compared to older patients.[4] Unlike the previous example, there was no unaccountable AI vendor involved. The system was developed and run by the UK's publicly funded health system. The developers' goals were aligned with ensuring good patient outcomes. This is a reminder that even when predictive algorithms could help in theory, it is hard to get the implementation right.

Predictive AI doesn't change much year-to-year. Not so with generative AI. Its capabilities and risks evolve quickly, and we regularly cover new developments on the AI Snake Oil newsletter (AISnakeOil.com). One trend we've seen is that the high cost of using large AI models, which we discussed in chapter 4, has fallen rapidly. Since the release of ChatGPT, the cost of generating a specific amount of output, at a specific level of output quality, has dropped over a hundredfold. This is because of improvements in both hardware and software efficiency.

Of course, this trend can't continue indefinitely, but it has already had big impacts. It has allowed AI companies to launch "reasoning" or "thinking" models that engage in an "inner monologue" before providing a response. The hope is to improve the quality of the outputs at the expense of speed and cost. The reasoning technique has improved AI's ability to write code and solve math problems. Both are domains where it is easier to check if a solution is correct than to generate a solution in the first place, so AI can get better by checking its own solutions (among many other techniques for enhancing accuracy). But improvements have been limited in domains where it is harder to verify correctness, or where correctness is not the only measure of quality, such as writing original prose.[5]

What all this means for AI's costs, energy consumption, and environmental impact is unclear. The cost of generating a unit of output is falling, but the amount of hidden output that tends to be generated for each response is increasing, and AI is being used for more complex tasks. Data centers, which is where the data storage and computation behind most of our digital activity happens, consume about 2 percent of global energy demand.[6] AI is a small fraction of that 2 percent, though it is growing quickly. So far, the most credible concern about AI and the environment is not its global footprint, but rather the fact that building new data centers tends to have highly localized impacts, putting heavy demands on specific regions and communities due to their high energy and water consumption.

How is AI being used for more complex tasks? Companies have built AI "agents" that can take actions based on user instructions, such as buying a product, identifying which emails need your attention, or finding and summarizing the available research on a given topic, sometimes taking an hour or more to complete a task. Many agents have been launched as consumer products. Agents exemplify a shift in how companies are trying to make money with advanced AI. While their earlier approach focused on simply building better models and letting users figure out what to do with them, agents tend to solve specific problems. This is a good thing. As we wrote in chapter 1, we want more AI applications that start working reliably and fade into the background. That can only happen when companies prioritize this goal.

Advances in AI capabilities have led to continued hype around Artificial General Intelligence. Many industry leaders have said, that agents will soon carry out a substantial portion of the economically viable tasks in the economy—for example, by serving as fully replaceable drop-ins for all remote workers.

Sam Altman has claimed that OpenAI now knows the roadmap to creating AGI and that the industry would achieve it during the second Trump administration (which ends in January 2029). Anthropic CEO Dario Amodei claimed they could get to AGI by 2027. Elon Musk predicted this would happen by 2026. Considering that CEOs have gradually been watering down what they mean by AGI, it's quite possible that they will declare victory in a few years. But let's set semantics aside and analyze what sort of impact AI is actually likely to have on labor.

AI agents are indeed useful for many tasks. But we don't think there will be drop-in AI replacements for human workers any time soon. That's because most jobs are bundles of many varied tasks. As agents start to automate or augment some tasks, job descriptions will naturally change to take advantage of AI abilities and refocus human effort on the tasks that can't yet be automated.

It is true that AI has had a big negative impact on jobs such as translating documents or transcribing speech into text. But these jobs are unusual in that they consist of a single, well-defined task (though there are exceptions—translating literature is very different from translating a product instruction leaflet). Thus, such services are often provided by freelancers rather than employees, a model that is particularly conducive to automation. When a job requires being embedded in an organization and having deep contextual awareness to complete many interlinked tasks, automation isn't nearly so simple.

Furthermore, many claims about AI agents being able to automate specific tasks are overblown. This includes many seemingly simple tasks, such as travel booking, as of 2025. This is because AI agents are *capable* of doing certain tasks but cannot do so *reliably*. If an AI travel agent books vacations to the wrong destination 10 percent of the time, it won't be a successful product. The AI industry was slow to recognize this gap, and it led to embarrassing failures of many products that launched with a great deal of buzz and hype.

Improving AI reliability will require sustained research and engineering; some of our own ongoing research is about laying the conceptual groundwork for this emerging field. Reliability is one of many factors that are critical in the real world but ignored in press releases that tout advances in AI capabilities. We expect improvements in real-world usefulness to be gradual, and not simply a matter of making models bigger or having them think for longer.

Just as the short-term benefits of AI are overstated, so are the risks. In 2024, over 60 countries held elections. AI-generated misinformation was one of the top concerns. The World Economic Forum claimed that "misinformation and disinformation is the most severe short-term risk the world faces" and that "AI is amplifying manipulated and distorted information that could destabilize societies."

But our analysis of AI use in the 2024 elections found that election deepfakes weren't nearly as disruptive as feared.[7] In elections worldwide, half of AI use wasn't deceptive—it was legitimate speech, such as satire. Even when AI was used deceptively, we found that the same content could have been created without AI, such as by hiring Photoshop experts, video editors, or voice actors. Such pieces of content are often called "cheap fakes" in contrast to deepfakes. Finally, the total number of AI deepfakes used in elections was low. WIRED's election deepfake tracker, which we used for our analysis, consisted of only 78 examples from across 60 countries that held elections in 2024. Another dataset, Spitting Images, documented a similar prevalence of election deepfakes. This was an order of magnitude lower than the total number of documented cheap fakes.[8]

To be clear, the legitimacy of elections worldwide is rightly an area of concern, and the world has been in a period of democratic decline since around 2010, with many more countries autocratizing than democratizing. But this is a social and political problem much more than it is a technological problem. Viewing AI as a culprit can be comforting because it suggests that there is a technological fix. But unfortunately this is only a distraction from the real issues.

We're not saying that AI deepfakes aren't a problem. One genuine concern is deepfake nudes of real people. Thousands of people have become victims of non-consensual deepfake nudes. Hundreds of thousands of images have been shared online. We suspect the real damage is much greater, as these numbers only reflect the incidents that were reported publicly.

Lawmakers around the world have gradually woken up to the need for legislation to curb the problem. Some countries, such as South Korea and the UK, have passed legislation on AI deepfakes. Others, such as Japan, use existing laws on harassment or cybercrime to target perpetrators. In the United States, federal legislation to address deepfake nudes has been introduced but has not yet been passed. (More broadly, federal policy relating to AI harms has been anemic, especially after the 2024 elections, with the Trump administration taking a much more industry-friendly approach than the Biden administration.) In the absence of federal legislation, many states have acted. In 2024, over 70 pieces of state legislation on AI deepfakes were enacted in over 20 U.S. states. One of the main areas of focus was the use of deepfakes to harass or defame people, such as by creating deepfake nudes.

Overall, AI policy worldwide has taken a more measured tone than a year ago. In the wake of ChatGPT's release, there was an urgent push to establish guardrails against many perceived risks of AI while grappling with existential threats to humanity. Now, with some distance, policymakers have had an opportunity to reflect more deliberately on which concerns were warranted and which may have been premature. Speculative existential risks have receded from the policy agenda. Other immediate threats, like the impact of AI-generated deepfakes on elections, have proven less disruptive than initially feared. Where urgent risks exist, such as with deepfake nudes, we see legal and regulatory approaches settling into place worldwide and at the U.S. state level, if not federally.

What remains is the need for a longer-term perspective that moves beyond AI's imminent threats to address deeper transformations to geopolitics, labor markets, and social institutions. Currently we lack an intellectual foundation to guide such sustained and coherent policymaking. Our next project aims to establish this foundation.

The project picks up where *AI Snake Oil* leaves off. We've acknowledged in this book that generative AI is already at a point where it is potentially useful to almost every knowledge worker, and that it is advancing rapidly. That raises an obvious question: What will the economic and social impact of advanced AI be in, say, twenty years? While we can't make precise predictions, it is useful to have some framework for thinking about the future of AI and its impacts—a framework that is grounded in evidence and not hype and speculation. In the final chapter, we discussed some aspects of this question, but didn't address it comprehensively.

Many AI industry leaders such as Dario Amodei predict a "country of geniuses in a data center," dramatically accelerating the pace of scientific research and technological innovation while also posing catastrophic risks if not properly controlled.[9] This assumption serves as the starting point for Amodei's analysis. This is not our view. Our skepticism of AI's existential risks (chapter 5) is based on a deeper disagreement: We think the usual conception of superintelligence is incoherent. Ironically, many proponents of AI utopia and dystopia resemble each other strikingly in core aspects of their thinking, viewing AI as an exception to well-worn patterns of technology development, adoption, and impact.

We view AI as *normal technology*. The term isn't meant to understate AI's potential impact: Even transformative general-purpose technologies such as electricity are "normal." What we mean is that we expect AI's adoption and impact to follow general patterns established by other transformative technologies. There is a lot of evidence that this is currently the path AI is on—based on decades of experience with predictive AI as well as early experience with generative AI. Our analysis is strongly guided by an expectation of continuity of AI's impacts between the past and the future.

We think that in most cases, simply attempting to replace human workers with AI will be an unviable strategy, *regardless* of AI's capability level. AI adoption will instead require gradual, uncertain, and often painful changes to organizational structures, worker practices, social norms, laws, and so forth, which tend to happen on a timescale of decades. (That said, if even 5 percent of jobs turn out to be relatively straightforward to automate in a short timeframe of, say, five years, it would cause significant hardship and potential social disruption, though far from cataclysmic.)

Social and economic transformations that unfold over decades aren't unprecedented. In fact, we've recently experienced one: the internet. It is a transformative technology in the sense that most cognitive tasks in the workforce today are mediated by it. But life goes on. Most job categories have survived relatively intact, though some have seen dramatic job losses and many new job categories have been created.

Crucially, the impact of the internet on economic indicators, notably the Gross Domestic Product, is hardly discernible. This is an obvious-in-retrospect insight behind the "normal technology" project: that a technology may qualitatively transform work without producing economic gains remotely resembling a utopia. Thus, many seemingly outlandish predictions—say, that AI might one day automate or at least mediate virtually every valuable task in today's economy—are in fact compatible with rather prosaic forecasts of impacts.

It's important to get this right. Viewing AI as normal technology results in drastically different predictions of the future than viewing it as a potential superintelligence. The two views also lead to policy implications that are often opposed to each other. The effects of policies, in areas ranging from industrial policy to AI safety, will have very different effects depending on which view better describes and predicts AI impacts. Decision makers in businesses also need to confront the tension between these views.

The good news that the normal technology view brings is that advanced AI isn't an imminent oncoming wave that we must all confront and somehow survive. The bad news is that there is no promised land waiting if only we manage this transition. While it will continue to bring many benefits, advanced AI will also continue to put many strains on our institutions and may stretch some of them to a breaking point. The work of reforming and rebuilding them will be painstaking and ongoing, but necessary and worthwhile. Let's settle in for the long haul.

ACKNOWLEDGMENTS

Writing *AI Snake Oil* would not have been possible without the invaluable contributions of many people. Hallie Stebbins gave us patient and priceless feedback on our chapter drafts several times over to help us turn a loose collection of chapters into a finished book draft. Matt Salganik not only inspired several chapters in the book with his trailblazing research but also offered invaluable feedback on chapter drafts. Melanie Mitchell, Molly Crockett, Serina Chang, Chris Bail, and two anonymous peer reviewers provided extensive comments on a draft that immeasurably improved our work.

We are grateful to our coauthors on various research projects that featured in or inspired the book. Collaborations and conversations with Solon Barocas, Katy Glenn Bass, Rishi Bommasani, Emily Cantrell, Peter Henderson, Daniel E. Ho, Jeremy Howard, Kevin Klyman, Mihir Kshirsagar, Seth Lazar, Percy Liang, Shayne Longpre, Kenny Peng, Ashwin Ramaswami, Hilke Schellmann, Ari Sen, Brandon Stewart, and Angelina Wang were crucial to the book. The research on account deletion presented in chapter 6 was conducted by Arvind in collaboration with our colleagues Nia M. Brazzell, Klaudia Jaźwińska, Orestis Papakyriakopoulos, and Angelina Wang.

We thank Julia Angwin, Xuechunzi Bai, Mitchell Baker, Kevin Bankston, Stella Biderman, Miranda Bogen, Deborah Bryant, Rumman Chowdhury, Peter Cihon, Justin Curl, Alex Engler, Ellie Evans, Andrew Gelman, Odd Erik Gundersen, Moritz Hardt, Dan Hendrycks, Mireille Hildebrandt, Kashmir Hill, Jake M. Hofman, Aspen Hopkins, Jessica Hullman, William Isaac, Ravi Iyer, Yacine Jernite, Melanie Kambadur, Daphne Keller, Zico Kolter, Michael A. Lones, Stefano Maffuli, Momin M. Malik, Nestor Maslej, Lisa Messeri, Priyanka Nanayakkara, Alondra Nelson, Hien Pham, Joelle Pineau, Russell A. Poldrack, Inioluwa Deborah Raji, Michael Roberts, Marta Serra-Garcia, Yonadav Shavit, Aviya Skowron, Victor Storchan, Jonathan Stray, Gilles Vandewiele, Betty Xiong, Cori Zarek, and Daniel Zhang for broader discussions that informed our work. We are grateful to Sanjeev Arora, Emily Bender, Timnit Gebru, and Colin Raffel for feedback on our newsletter posts and talks.

Finally, we are grateful to our colleagues, present and former, at the Princeton Center for Information Technology Policy, who helped create the atmosphere for productive collaborations and influenced our writing in various parts of the book: Archana Ahlawat, Shazeda Ahmed, Jordan Brensinger, Jean Butcher, Dan Calacci, Tithi Chattopadhyay, Laura Cummings-Abdo, Amrit Daswaney, Shreyas Gandlur, Lucy He, Ben Kaiser, Anne Kolhbrenner, Aleksandra Korolova, Amna Liaqat, Eli Lucherini, Surya Mattu, Jonathan Mayer, Jakob Mökander, Andrés Monroy-Hernández, Nitya Nadgir, Varun Rao, Karen Rouse, Benedikt Ströbl, Matthew Sun, Ross Teixeira, Christelle Tessono, Mona Wang, Elizabeth Watkins, Amy Winecoff, and Madelyne Xiao.

REFERENCES

Chapter 1. Introduction

1. Heaven WD. "The Inside Story of How ChatGPT Was Built from the People Who Made It." *MIT Technology Review*. March 3, 2023. https://www.technologyreview.com/2023/03/03/1069311/inside-story-oral-history-how-chatgpt-built-openai/

2. Ptacek TH. "I'm sorry, I simply cannot be cynical about a technology that can accomplish this." X (formerly Twitter). December 1, 2022. https://twitter.com/tqbf/status/1598513757805858820?lang=en

3. Hu K. "ChatGPT Sets Record for Fastest-Growing User Base—Analyst Note." Reuters. February 2, 2023. https://www.reuters.com/technology/chatgpt-sets-record-fastest-growing-user-base-analyst-note-2023-02-01/

4. DeGeurin M. "Why Google Isn't Rushing Forward with AI Chatbots." Gizmodo. December 14, 2022. https://gizmodo.com/lamda-google-ai-chatgpt-openai-1849892728

5. Macintosh B [@bmac_astro]. "Speaking as someone who imaged an exoplanet 14 years before JWST was launched, it feels like you should find a better example?" X (formerly Twitter). February 7, 2023. https://twitter.com/bmac_astro/status/1623136549524353024

6. Wittenstein J. "Bard AI Chatbot Just Cost Google $100 Billion." *Time*. February 9, 2023. https://time.com/6254226/alphabet-google-bard-100-billion-ai-error/

7. Christian J. "CNET Sister Site Restarts AI Articles, Immediately Publishes Idiotic Error." Futurism. Updated February 1, 2023. https://futurism.com/cnet-bankrate-restarts-ai-articles

8. Cole S. "'Life or Death:' AI-Generated Mushroom Foraging Books Are All over Amazon." 404 Media. August 29, 2023. https://www.404media.co/ai-generated-mushroom-foraging-books-amazon/

9. Brittain B. "OpenAI Asks Court to Trim Authors' Copyright Lawsuits." Reuters. August 29, 2023. https://www.reuters.com/legal/litigation/openai-asks-court-trim-authors-copyright-lawsuits-2023-08-29/

10. Valyaeva A. "AI Image Statistics: How Much Content Was Created by AI." *Insight* (blog). Everypixel Journal. August 15, 2023. https://journal.everypixel.com/ai-image-statistics

11. Kapoor S, Narayanan A. "How to Prepare for the Deluge of Generative AI on Social Media." Kn First Amend Inst. June 16, 2023. http://knightcolumbia.org/content/how-to-prepare-for-the-deluge-of-generative-ai-on-social-media

12. NVIDIA Game Developer. "NVIDIA ACE for Games Sparks Life into Virtual Characters with Generative AI." YouTube video, 2:02. May 28, 2023. https://www.youtube.com/watch?v=nAEQdF3JAJo

13. Dalton A. "AI Is the Wild Card in Hollywood's Strikes. Here's an Explanation of Its Unsettling Role." AP News. July 21, 2023. https://apnews.com/article/artificial-intelligence-hollywood-strikes-explained-writers-actors-e872bd63ab52c3ea9f7d6e825240a202

14. Crawford K. *Atlas of AI: Power, Politics, and the Planetary Costs of Artificial Intelligence*. New Haven: Yale University Press; 2021.

15. Narayanan A. "Students Are Acing Their Homework by Turning in Machine-Generated Essays. Good." AI Snake Oil. October 21, 2022. https://www.aisnakeoil.com/p/students-are-acing-their-homework

16. Raji ID, Kumar IE, Horowitz A, Selbst A. "The Fallacy of AI Functionality." In *2022 ACM Conference on Fairness, Accountability, and Transparency*. Seoul Republic of Korea: ACM; 2022. p. 959–72. https://dl.acm.org/doi/10.1145/3531146.3533158

17. Ross C, Herman B. "Denied by AI: How Medicare Advantage Plans Use Algorithms to Cut Off Care for Seniors in Need." STAT. March 13, 2023. https://www.statnews.com/2023/03/13/medicare-advantage-plans-denial-artificial-intelligence/

18. Varner M, Sankin A. "Suckers List: How Allstate's Secret Auto Insurance Algorithm Squeezes Big Spenders." The Markup. February 25, 2020. https://themarkup.org/allstates-algorithm/2020/02/25/car-insurance-suckers-list

19. Robinson DG. *Voices in the Code: A Story about People, Their Values, and the Algorithm They Made*. New York: Russell Sage Foundation; 2022.

20. Marcus G. "Face It, Self-Driving Cars Still Haven't Earned Their Stripes." *Marcus on AI* (blog). August 19, 2023. https://garymarcus.substack.com/p/face-it-self-driving-cars-still-havent

21. Ryan-Mosley T. "The New Lawsuit That Shows Facial Recognition Is Officially a Civil Rights Issue." *MIT Technology Review*. April 14, 2021. https://www.technologyreview.com/2021/04/14/1022676/robert-williams-facial-recognition-lawsuit-aclu-detroit-police/

22. "Facial Recognition Tool Led to Mistaken Arrest, Lawyer Says." AP News. January 2, 2023. https://apnews.com/article/technology-louisiana-baton-rouge-new-orleans-crime-50e1ea591aed6cf14d248096958dccc4

23. Hill K. "Eight Months Pregnant and Arrested after False Facial Recognition Match." *New York Times*. August 6, 2023. https://www.nytimes.com/2023/08/06/business/facial-recognition-false-arrest.html

24. Cipriano A. "Facial Recognition Now Used in over 1,800 Police Agencies: Report." The Crime Report. April 7, 2021. https://thecrimereport.org/2021/04/07/facial-recognition-now-used-in-over-1800-police-agencies-report/

25. Crumpler W. "How Accurate Are Facial Recognition Systems—and Why Does It Matter?" *Strategic Technologies Blog*. CSIS. April 14, 2020. https://www.csis.org/blogs/strategic-technologies-blog/how-accurate-are-facial-recognition-systems-and-why-does-it

26. Buolamwini J, Gebru T, Raynham H, Raji D, Zuckerman E. "Gender Shades." MIT Media Lab. https://www.media.mit.edu/projects/gender-shades/overview/. Accessed February 15, 2024.

27. Buolamwini J, Gebru T. "Gender Shades: Intersectional Accuracy Disparities in Commercial Gender Classification." In: *Proceedings of the 1st Conference on Fairness, Accountability and Transparency*. PMLR; 2018. pp. 77–91. https://proceedings.mlr.press/v81/buolamwini18a.html

28. Hill K. *Your Face Belongs to Us: A Secretive Startup's Quest to End Privacy as We Know It*. New York: Random House; 2023.

29. "Russia: Police Target Peaceful Protesters Identified Using Facial Recognition Technology." Amnesty International. April 27, 2021. https://www.amnesty.org/en/latest/press-release/2021/04/russia-police-target-peaceful-protesters-identified-using-facial-recognition-technology/

30. Hill K, Kilgannon C. "Madison Square Garden Uses Facial Recognition to Ban Its Owner's Enemies." *New York Times*. December 22, 2022. https://www.nytimes.com/2022/12/22/nyregion/madison-square-garden-facial-recognition.html

31. Brewster T. "Exclusive: DHS Used Clearview AI Facial Recognition in Thousands of Child Exploitation Cold Cases." *Forbes*. August 7, 2023. https://www.forbes.com/sites/thomasbrewster/2023/08/07/dhs-ai-facial-recognition-solving-child-exploitation-cold-cases/

32. "Rite Aid Corporation, FTC v." Cases and Proceedings. Federal Trade Commission. 2023. https://www.ftc.gov/legal-library/browse/cases-proceedings/2023190-rite-aid-corporation-ftc-v

33. Kaltheuner F., ed. *Fake AI*. Manchester, UK: Meatspace Press; 2021. https://shop.meatspacepress.com/products/fake-ai-e-book

34. Lazer D, Kennedy R, King G, Vespignani A. "The Parable of Google Flu: Traps in Big Data Analysis." *Science* 343, no. 6176 (March 2014):11203–5.

35. Sculley D, Holt G, Golovin D, Davydov E, Phillips T, Ebner D, et al. "Hidden Technical Debt in Machine Learning Systems." In: *Advances in Neural Information Processing Systems*. Montreal: Curran Associates; 2015. https://papers.nips.cc/paper_files/paper/2015/hash/86df7dcfd896fcaf2674f757a2463eba-Abstract.html

36. Merritt SH, Gaffuri K, Zak PJ. "Accurately Predicting Hit Songs Using Neurophysiology and Machine Learning." *Front Artif Intell*. 6 (2023). https://www.frontiersin.org/articles/10.3389/frai.2023.1154663

37. Bushwick S, Tu L. "Here's How AI Can Predict Hit Songs with Frightening Accuracy." *Scientific American*. July 28, 2023. https://www.scientificamerican.com/podcast/episode/heres-how-ai-can-predict-hit-songs-with-frightening-accuracy/

38. Heath R. "Neuro-forecasting the Next No. 1 Song." Axios. June 27, 2023. https://www.axios.com/2023/06/27/ai-predicts-hits-number-one-songs

39. Kapoor S, Narayanan A. "Leakage and the Reproducibility Crisis in Machine-Learning-Based Science." *Patterns* 4, no. 9 (September 2023):100804.

40. Roberts M, Driggs D, Thorpe M, Gilbey J, Yeung M, Ursprung S, et al. "Common Pitfalls and Recommendations for Using Machine Learning to Detect and Prognosticate for COVID-19 Using Chest Radiographs and CT Scans." *Nat Mach Intell*. 3, no 3 (March 2021):199–217.

41. Serra-Garcia M, Gneezy U. "Nonreplicable Publications Are Cited More Than Replicable Ones." *Sci Adv*. 7, no. 21 (May 2021):eabd1705.

42. Kapoor S, Cantrell E, Peng K, Pham TH, Bail CA, Gundersen OE, et al. "REFORMS: Reporting Standards for Machine Learning Based Science." arXiv. Revised September 19, 2023. http://arxiv.org/abs/2308.07832

43. Wang A, Kapoor S, Barocas S, Narayanan A. "Against Predictive Optimization: On the Legitimacy of Decision-Making Algorithms That Optimize Predictive Accuracy." In: *2023 ACM Conference on Fairness, Accountability, and Transparency*. Chicago, IL: ACM; 2023. p. 626–626. https://dl.acm.org/doi/10.1145/3593013.3594030

44. Atleson M. "Keep Your AI Claims in Check." *Business blog*. Federal Trade Commission. February 27, 2023. https://www.ftc.gov/business-guidance/blog/2023/02/keep-your-ai-claims-check

45. Levy S. "Blake Lemoine Says Google's LaMDA AI Faces 'Bigotry.'" *Wired*. June 17, 2022. https://www.wired.com/story/blake-lemoine-google-lamda-ai-bigotry/

46. Pulitzer Center. "AI Accountability Fellowships." Pulitzer Center. 2023. https://pulitzercenter.org/grants-fellowships/opportunities-journalists/ai-accountability-fellowships

47. Sen A, Bennett DK. "Tracked: How Colleges Use AI to Monitor Student Protests." *Dallas Morning News*. September 20, 2022. https://interactives.dallasnews.com/2022/social-sentinel/

48. Narayanan A, Kapoor S. AI Snake Oil. https://www.aisnakeoil.com/

49. Yiu E, Kosoy E, Gopnik A. "Transmission versus Truth, Imitation versus Innovation: What Children Can Do That Large Language and Language-and-Vision Models Cannot (Yet)." *Perspect Psychol Sci J Assoc Psychol Sci.* (October 2023):17456916231201401.

50. Li H, Vincent N, Chancellor S, Hecht B. "The Dimensions of Data Labor: A Road Map for Researchers, Activists, and Policymakers to Empower Data Producers." In: *Proceedings of the 2023 ACM Conference on Fairness, Accountability, and Transparency.* New York: ACM; 2023. p. 1151–61. https://dl.acm.org/doi/10.1145/3593013.3594070

Chapter 2. How Predictive AI Goes Wrong

1. Svrluga S. "University President Allegedly Says Struggling Freshmen Are Bunnies That Should Be Drowned." *Washington Post.* January 19, 2016. https://www.washingtonpost.com/news/grade-point/wp/2016/01/19/university-president-allegedly-says-struggling-freshmen-are-bunnies-that-should-be-drowned-that-a-glock-should-be-put-to-their-heads/

2. Feathers T. "Major Universities Are Using Race as a 'High Impact Predictor' of Student Success." The Markup. March 2, 2021. https://themarkup.org/machine-learning/2021/03/02/major-universities-are-using-race-as-a-high-impact-predictor-of-student-success

3. Waldman A. "Power, Process, and Automated Decision-Making." *Fordham Law Rev.* 88, no. 2 (November 2019):613.

4. Artificial Intelligence Incident Database. https://incidentdatabase.ai/

5. AI, Algorithmic, and Automation Incidents and Controversies Repository. https://www.aiaaic.org/aiaaic-repository

6. Pasquale F. *The Black Box Society: The Secret Algorithms That Control Money and Information.* Cambridge, MA: Harvard University Press; 2015. https://www.degruyter.com/document/doi/10.4159/harvard.9780674736061/html

7. "Advancing Public Health Interventions to Address the Harms of the Carceral System." Policy Statement Database. APHA. October 26, 2021. https://www.apha.org/Policies-and-Advocacy/Public-Health-Policy-Statements/Policy-Database/2022/01/07/Advancing-Public-Health-Interventions-to-Address-the-Harms-of-the-Carceral-System

8. Rabuy B, Kopf D. "Detaining the Poor: How Money Bail Perpetuates an Endless Cycle of Poverty and Jail Time." Prison Policy Initiative. May 10, 2016. https://www.prisonpolicy.org/reports/incomejails.html

9. Subramanian R et al. *Incarceration's Front Door: The Misuse of Jails in America.* New York: Vera Institute of Justice; 2015. https://www.vera.org/downloads/publications/incarcerations-front-door-summary.pdf

10. Kang-Brown J, Montagnet C Heiss J. *People in Jail and Prison in Spring 2021*. New York: Vera Institute of Justice; 2021. https://www.vera.org/downloads/publications/people-in-jail-and-prison-in-spring-2021.pdf

11. Angwin J (ProPublica), contributor. *Sample COMPAS Risk Assessment: COMPAS "CORE."* DocumentCloud; 2011. https://www.documentcloud.org/documents/2702103-Sample-Risk-Assessment-COMPAS-CORE

12. Northpointe, Inc. *Practitioner's Guide to COMPAS Core*. April 4, 2019. http://www.equivant.com/wp-content/uploads/Practitioners-Guide-to-COMPAS-Core-040419.pdf

13. Taulli T. "Upstart: Can AI Kill the FICO Score?" *Forbes*. August 13, 2021. https://www.forbes.com/sites/tomtaulli/2021/08/13/upstart-can-ai-kill-the-fico-score/

14. Lum K, Isaac W. "To Predict and Serve?" *Significance* 13, no. 5 (2016):14–9.

15. Wang A, Kapoor S, Barocas S, Narayanan A. "Against Predictive Optimization: On the Legitimacy of Decision-Making Algorithms That Optimize Predictive Accuracy." In: *2023 ACM Conference on Fairness, Accountability, and Transparency*. Chicago, IL: ACM; 2023. p. 626. https://dl.acm.org/doi/10.1145/3593013.3594030

16. Caruana R, Lou Y, Gehrke J, Koch P, Sturm M, Elhadad N. "Intelligible Models for HealthCare: Predicting Pneumonia Risk and Hospital 30-Day Readmission." In: *Proceedings of the 21st ACM SIGKDD International Conference on Knowledge Discovery and Data Mining*. Sydney, NSW Australia: ACM; 2015. p. 1721–30. https://dl.acm.org/doi/10.1145/2783258.2788613

17. Cooper GF, Aliferis CF, Ambrosino R, Aronis J, Buchanan BG, Caruana R, et al. "An Evaluation of Machine-Learning Methods for Predicting Pneumonia Mortality." *Artif Intell Med* 9, no. 2 (February 1997):107–38.

18. Ye C, Fu T, Hao S, Zhang Y, Wang O, Jin B, et al. "Prediction of Incident Hypertension within the Next Year: Prospective Study Using Statewide Electronic Health Records and Machine Learning." *J Med Internet Res* 20, no. 1 (January 2018):e22.

19. Filho AC, Batista AFDM, Santos HG dos. "Data Leakage in Health Outcomes Prediction with Machine Learning. Comment on 'Prediction of Incident Hypertension within the Next Year: Prospective Study Using Statewide Electronic Health Records and Machine Learning.'" *J Med Internet Res* 23, no. 2 (February 2021):e10969.

20. Fuller JB, Raman M, Sage-Gavin E, Hines K. *Hidden Workers: Untapped Talent*. Cambridge, MA: Harvard Business School; 2021. https://www.hbs.edu/managing-the-future-of-work/Documents/research/hiddenworkers09032021.pdf

21. Burrell J. "How the Machine 'Thinks': Understanding Opacity in Machine Learning Algorithms." *Big Data Soc* 3, no. 1 (June 2016):2053951715622512.

22. Schellmann H. "Finding It Hard to Get a New Job? Robot Recruiters Might Be to Blame." *Guardian*. May 11, 2022. https://www.theguardian.com/us-news/2022/may/11/artitifical-intelligence-job-applications-screen-robot-recruiters

23. Harwell D. "A Face-Scanning Algorithm Increasingly Decides Whether You Deserve the Job." *Washington Post*. October 22, 2019. https://www.washingtonpost.com/technology/2019/10/22/ai-hiring-face-scanning-algorithm-increasingly-decides-whether-you-deserve-job/

24. Harlan E, Schnuck O. "Objective or Biased." BR 24. February 16, 2021. https://interaktiv.br.de/ki-bewerbung/en/

25. Rhea A, Markey K, D'Arinzo L, Schellmann H, Sloane M, Squires P, et al. "Resume Format, LinkedIn URLs and Other Unexpected Influences on AI Personality Prediction in Hiring: Results of an Audit." In: *Proceedings of the 2022 AAAI/ACM Conference on AI, Ethics, and Society*. Oxford: ACM; 2022. p. 572–87. https://dl.acm.org/doi/10.1145/3514094.3534189

26. Geiger G. "How a Discriminatory Algorithm Wrongly Accused Thousands of Families of Fraud." Vice. March 1, 2021. https://www.vice.com/en/article/jgq35d/how-a-discriminatory-algorithm-wrongly-accused-thousands-of-families-of-fraud

27. Heikkilä M. "Dutch Scandal Serves as a Warning for Europe over Risks of Using Algorithms." POLITICO. March 29, 2022. https://www.politico.eu/article/dutch-scandal-serves-as-a-warning-for-europe-over-risks-of-using-algorithms/

28. Jones S. "Many Caribbean Dutch Victims of Benefits Scandal." Caribbean Network. February 1, 2021. https://caribischnetwerk.ntr.nl/2021/02/01/veel-caribische-nederlanders-slachtoffer-toeslagenaffaire/

29. Autoriteit Persoonsgegevens. "Tax Authorities Fine for FSV Blacklist." April 12, 2022. https://autoriteitpersoonsgegevens.nl/actueel/boete-belastingdienst-voor-zwarte-lijst-fsv

30. Wykstra S, Undark. "It Was Supposed to Detect Fraud. It Wrongfully Accused Thousands Instead." *Atlantic*. 2020. https://www.theatlantic.com/technology/archive/2020/06/michigan-unemployment-fraud-automation/612721/

31. Pearson J. "The Story of How the Australian Government Screwed Its Most Vulnerable People." Vice. August 24, 2020. https://www.vice.com/en/article/y3zkgb/the-story-of-how-the-australian-government-screwed-its-most-vulnerable-people-v27n3

32. Martineau P. "Toronto Tapped Artificial Intelligence to Warn Swimmers. The Experiment Failed." The Information. November 4, 2022. https://www.theinformation.com/articles/when-artificial-intelligence-isnt-smarter

33. Mole B. "UnitedHealth Uses AI Model with 90% Error Rate to Deny Care, Lawsuit Alleges." Ars Technica. November 16, 2023. https://arstechnica.com/health/2023/11/ai-with-90-error-rate-forces-elderly-out-of-rehab-nursing-homes-suit-claims/

34. Parasuraman R, Manzey DH. "Complacency and Bias in Human Use of Automation: An Attentional Integration." *Hum Factors* 52, no. 3 (June 2010):381–410.

35. Goel S, Shroff R, Skeem J, Slobogin C. "The Accuracy, Equity, and Jurisprudence of Criminal Risk Assessment." In: *Research Handbook on Big Data Law*, ed.

Roland Vogel. Cheltenham, UK: Edward Elgar Publishing; 2021. p. 9–28. https://www.elgaronline.com/edcollchap/edcoll/9781788972819/9781788972819.00007.xml

36. Corey E. "How a Tool to Help Judges May Be Leading Them Astray." *The Appeal*. August 8, 2019. https://theappeal.org/how-a-tool-to-help-judges-may-be-leading-them-astray/

37. Chouldechova A, Benavides-Prado D, Fialko O, Vaithianathan R. "A Case Study of Algorithm-Assisted Decision Making in Child Maltreatment Hotline Screening Decisions." In: *Proceedings of the 1st Conference on Fairness, Accountability and Transparency*. PMLR; 2018. p. 134–48. https://proceedings.mlr.press/v81/chouldechova18a.html

38. Abdurahman JK. "Birthing Predictions of Premature Death." *Logic(s) Magazine*. August 22, 2022. https://logicmag.io/home/birthing-predictions-of-premature-death/

39. Obermeyer Z, Powers B, Vogeli C, Mullainathan S. "Dissecting Racial Bias in an Algorithm Used to Manage the Health of Populations." *Science* 366, no. 6464 (October 2019):447–53.

40. Wolford G, Miller MB, Gazzaniga M. "The Left Hemisphere's Role in Hypothesis Formation." *J Neurosci* 20, no. 6 (March 2000):RC64.

41. Jenkins HM, Ward WC. "Judgment of Contingency between Responses and Outcomes." *Psychol Monogr Gen Appl* 79, no. 1 (1965):1–17.

42. Nix E. "'Dewey Defeats Truman': The Election Upset behind the Photo." *History*. Updated November 2, 2020. https://www.history.com/news/dewey-defeats-truman-election-headline-gaffe

43. Wohlsen M. "I Just Want Nate Silver to Tell Me Everything's Going to Be Fine." *Wired*. October 16, 2016. https://www.wired.com/2016/10/just-want-nate-silver-tell-everythings-going-fine/

44. Bueno N, Nunes F, Zucco C. "Benefits by Luck: A Study of Lotteries as a Selection Method for Government Programs." Available at SSRN (April 2023). https://papers.ssrn.com/abstract=4411082

Chapter 3. Why Can't AI Predict the Future?

1. Rees A. "The History of Predicting the Future." *Wired*. December 27, 2021. https://www.wired.com/story/history-predicting-future/

2. "Diaper-Beer Syndrome." *Forbes*. April 6, 1998. https://www.forbes.com/forbes/1998/0406/6107128a.html

3. Anderson C. "The End of Theory: The Data Deluge Makes the Scientific Method Obsolete." *Wired*. June 23, 2008. https://www.wired.com/2008/06/pb-theory/

4. Narayanan A, Salganik MJ. "Limits to Prediction." Fall 2020. https://msalganik.github.io/cos597E-soc555_f2020/

5. Romeo N. "Ancient Device for Determining Taxes Discovered in Egypt." *National Geographic.* May 18, 2016. https://www.nationalgeographic.com/history/article/160517-nilometer-discovered-ancient-egypt-nile-river-archaeology

6. "Weather Forecasting through the Ages." NASA Earth Observatory. February 25, 2002. https://earthobservatory.nasa.gov/features/WxForecasting/wx2.php

7. Lorenz EN. "Deterministic Nonperiodic Flow." *J Atmospheric Sci* 20, no. 2 (March 1963):130–41.

8. Chang K, Edward N. "Lorenz, a Meteorologist and a Father of Chaos Theory, Dies at 90." *New York Times.* April 17, 2008. https://www.nytimes.com/2008/04/17/us/17lorenz.html

9. "Butterflies, Tornadoes, and Time Travel." APSNews. June 2004. http://www.aps.org/publications/apsnews/200406/butterfly-effect.cfm

10. Bauer P, Thorpe A, Brunet G. "The Quiet Revolution of Numerical Weather Prediction." *Nature* 525, no. 7567 (September 2015):47–55.

11. Forrester JW. "System Dynamics and the Lessons of 35 Years." In: *A Systems-Based Approach to Policymaking*, ed. KB Greene. Boston, MA: Springer US; 1993. p. 199–240. https://doi.org/10.1007/978-1-4615-3226-2_7

12. Lepore J. "How the Simulmatics Corporation Invented the Future." *New Yorker.* July 27, 2020. https://www.newyorker.com/magazine/2020/08/03/how-the-simulmatics-corporation-invented-the-future

13. De Sola Pool I, Abelson R. "The Simulmatics Project." *Public Opin Q* 25, no. 2 (January 1961):167–83.

14. Paul T. "When Did Credit Scores Start? A Brief Look at the Long History behind Credit Reporting." CNBC. Updated January 31, 2023. https://www.cnbc.com/select/when-did-credit-scores-start/

15. Ochigame R. "The Long History of Algorithmic Fairness." Phenomenal World. January 30, 2020. https://www.phenomenalworld.org/analysis/long-history-algorithmic-fairness/

16. Reicin E. "Council Post: AI Can Be a Force for Good in Recruiting and Hiring New Employees." *Forbes.* November 16, 2021. https://www.forbes.com/sites/forbesnonprofitcouncil/2021/11/16/ai-can-be-a-force-for-good-in-recruiting-and-hiring-new-employees/

17. Lau J. "Google Maps 101: How AI Helps Predict Traffic and Determine Routes." *The Keyword* (blog). Google. September 3, 2020. https://blog.google/products/maps/google-maps-101-how-ai-helps-predict-traffic-and-determine-routes/

18. "Traffic Prediction for Retail Labor Forecasting & More." SenSource. https://sensourceinc.com/vea-software/forecasting/. Accessed February 16, 2024.

19. Earthquake Hazards Program. "Introduction to the National Seismic Hazard Maps." U.S. Geological Survey. March 9, 2022. https://www.usgs.gov/programs/earthquake-hazards/science/introduction-national-seismic-hazard-maps

20. "Can You Predict Earthquakes?" U.S. Geological Survey. https://www.usgs.gov/faqs/can-you-predict-earthquakes. Accessed February 16, 2024.

21. Hong S-ha. "Prediction as Extraction of Discretion." *Big Data Soc* 10, no. 1 (January 2023):20539517231171053.

22. Hofman JM, Sharma A, Watts DJ. "Prediction and Explanation in Social Systems." *Science* 355 no. 6324 (February 2017):486–8.

23. Scott SB, Rhoades GK, Stanley SM, Allen ES, Markman HJ. "Reasons for Divorce and Recollections of Premarital Intervention: Implications for Improving Relationship Education." *Couple Fam Psychol* 2, no. 2 (June 2013):131–45.

24. Heyman RE, Slep AMS. "The Hazards of Predicting Divorce without Cross-validation." *J Marriage Fam* 63, no. 2 (2001):473–9.

25. Salganik MJ, Lundberg I, Kindel AT, Ahearn CE, Al-Ghoneim K, Almaatouq A, et al. "Measuring the Predictability of Life Outcomes with a Scientific Mass Collaboration." *Proc Natl Acad Sci* 117, no. 15 (April 2020):8398–403.

26. "Download ImageNet Data." Imagenet. 2020. https://www.image-net.org/download.php

27. Narayanan A, Salganik MJ. "Limits to Prediction": pre-read. Fall 2020.

28. Lundberg I, Brown-Weinstock R, Clampet-Lundquist S, Pachman S, Nelson TJ, Yang V, et al. "The Origins of Unpredictability in Life Trajectory Prediction Tasks." arXiv. October 19, 2023. http://arxiv.org/abs/2310.12871

29. van der Laan J, de Jonge E, Das M, Te Riele S, Emery T. "A Whole Population Network and Its Application for the Social Sciences." *Eur Sociol Rev* 39, no. 1 (February 2023):145–60.

30. "SICSS-ODISSEI Schedule & Materials." SICSS. 2023. https://sicss.io/2023/odissei/schedule

31. Farahany NA. *The Battle for Your Brain: Defending the Right to Think Freely in the Age of Neurotechnology*. New York: St. Martin's Press; 2023.

32. Wang A, Kapoor S, Barocas S, Narayanan A. "Against Predictive Optimization: On the Legitimacy of Decision-Making Algorithms That Optimize Predictive Accuracy." In: *2023 ACM Conference on Fairness, Accountability, and Transparency*. Chicago, IL: ACM; 2023. p. 62. https://dl.acm.org/doi/10.1145/3593013.3594030

33. Northpointe. *Practitioner's Guide to COMPAS Core*. April 4, 2019. http://www.equivant.com/wp-content/uploads/Practitioners-Guide-to-COMPAS-Core-040419.pdf

34. Angwin J, Larson J, Mattu S, Kirchner L. "Machine Bias." ProPublica. May 23, 2016. https://www.propublica.org/article/machine-bias-risk-assessments-in-criminal-sentencing

35. Larson J, Mattu S, Kirchner L, Angwin J. "How We Analyzed the COMPAS Recidivism Algorithm." ProPublica. May 23, 2016. https://www.propublica.org/article/how-we-analyzed-the-compas-recidivism-algorithm

36. Pierson E, Simoiu C, Overgoor J, Corbett-Davies S, Jenson D, Shoemaker A, et al. "A Large-Scale Analysis of Racial Disparities in Police Stops across the United States." *Nat Hum Behav* 4, no. 7 (July 2020):736–45.

37. Dressel J, Farid H. "The Accuracy, Fairness, and Limits of Predicting Recidivism." *Sci Adv* 4, no. 1 (2018):eaao5580.

38. Steinberg L, Scott ES. "Less Guilty by Reason of Adolescence: Developmental Immaturity, Diminished Responsibility, and the Juvenile Death Penalty." *Am Psychol* 58, no. 12 (December 2003):1009–18.

39. "Why Luck Is the Silent Partner of Success." Knowledge at Wharton. October 20, 2017. https://knowledge.wharton.upenn.edu/article/how-luck-is-the-silent-partner-of-success/

40. Mlodinow L. *The Drunkard's Walk: How Randomness Rules Our Lives.* Reprint edition. New York: Vintage; 2009.

41. "20 Years Ago: Tom Brady Replaced an Injured Drew Bledsoe, Changing the Patriots Franchise Forever." CBS Boston. September 23, 2021. https://www.cbsnews.com/boston/news/20-years-ago-tom-brady-replaced-drew-bledsoe-changing-patriots-franchise-forever-mo-lewis-nfl-bill-belichick/

42. Chmielewski DC. "Disney Expects $200-Million Loss on 'John Carter.'" *Los Angeles Times.* March 20, 2012. https://www.latimes.com/entertainment/la-xpm-2012-mar-20-la-fi-ct-disney-write-down-20120320-story.html

43. D'Alessandro A. "'Strange World' to Lose $147M: Why Theatrical Was Best Decision for Doomed Toon—Not Disney+—as Bob Iger Takes Over CEO from Bob Chapek." Deadline. November 27, 2022. https://deadline.com/2022/11/strange-world-bombs-box-office-disney-glass-onion-bob-iger-bob-chapek-1235182222/

44. Whitbrook J. "How a *Trigun* Stan Made a 2019 Sci-Fi Novel a Hit on Amazon." Gizmodo. May 10, 2023. https://gizmodo.com/this-is-how-you-lose-the-time-war-trigun-twitter-amazon-1850424312

45. Bol T, de Vaan M, van de Rijt A. "The Matthew Effect in Science Funding." *Proc Natl Acad Sci* 115, no. 19 (May 2018):4887–90.

46. Goel S, Anderson A, Hofman J, Watts DJ. "The Structural Virality of Online Diffusion." *Manag Sci* 62, no. 1 (January 2016):180–96.

47. Guinaudeau B, Munger K, Votta F. "Fifteen Seconds of Fame: TikTok and the Supply Side of Social Video." *Comput Commun Res* 4, no. 2 (October 2022): 463–85.

48. Kleinman Z. "'Charlie Bit My Finger' Video to Be Taken Off YouTube after Selling for £500,000." BBC. May 24, 2021. https://www.bbc.com/news/newsbeat-57227290

49. Martin T, Hofman JM, Sharma A, Anderson A, Watts DJ. "Exploring Limits to Prediction in Complex Social Systems." In: *Proceedings of the 25th International Conference on World Wide Web*. Republic and Canton of Geneva, CHE: International World Wide Web Conferences Steering Committee; 2016. p. 683–94. https://dl.acm.org/doi/10.1145/2872427.2883001

50. Zukin M. "Why TikTok Stars Will Survive No Matter What." *Variety*. August 4, 2020. https://variety.com/2020/digital/news/tiktok-stars-charli-damelio-noah-schnapp-jalaiah-harmon-1234723975/

51. Ronson J. "How One Stupid Tweet Blew Up Justine Sacco's Life." *New York Times*. February 12, 2015. https://www.nytimes.com/2015/02/15/magazine/how-one-stupid-tweet-ruined-justine-saccos-life.html

52. Robertson CE, Pröllochs N, Schwarzenegger K, Pärnamets P, Van Bavel JJ, Feuerriegel S. "Negativity Drives Online News Consumption." *Nat Hum Behav* 7, no. 5 (May 2023):812–22.

53. Heltzel G, Laurin K. "Polarization in America: Two Possible Futures." *Curr Opin Behav Sci* 34 (August 2020):179–84.

54. Zachariah RA, Sharma S, Kumar V. "Systematic Review of Passenger Demand Forecasting in Aviation Industry." *Multimed Tools Appl* 1 (May 2023):1–37.

55. Turchin P. *Ages of Discord: A Structural-Demographic Analysis of American History*. Chaplin, CT: Beresta Books; 2016.

56. Chancellor E. "The Bubble in Predicting the End of the World." Reuters. December 1, 2022. https://www.reuters.com/breakingviews/global-markets-breakingviews-2022-12-01/

57. Tetlock PE. *Expert Political Judgment: How Good Is It? How Can We Know?* New edition. Princeton, NJ: Princeton University Press; 2017.

58. Turchin P. "Fitting Dynamic Regression Models to Seshat Data." *Cliodynamics* 9, no. 1 (June 2018). https://escholarship.org/uc/item/99x6r11m

59. "About CDC's Flu Forecasting Efforts." CDC. last reviewed October 6, 2023. https://www.cdc.gov/flu/weekly/flusight/how-flu-forecasting.htm

60. "FluSight: Flu Forecasting for Influenza Prevention and Control." CDC. Last reviewed October 10, 2023. https://www.cdc.gov/flu/weekly/flusight/index.html

61. Osterholm MT. "The Fog of Pandemic Planning." CIDRAP. University of Minnesota. January 31, 2007. https://www.cidrap.umn.edu/business-preparedness/fog-pandemic-planning

62. Global Preparedness Monitoring Board. *A World at Risk: Annual Report 2019*. Geneva: World Health Organization. September 2019. https://www.gpmb.org/reports/annual-report-2019

63. Gates B. "The Next Outbreak? We're Not Ready." Filmed March 2015 in Vancouver, BC. TED video, 8:24. https://www.ted.com/talks/bill_gates_the_next_outbreak_we_re_not_ready?language=dz

64. Binny RN, Lustig A, Hendy SC, Maclaren OJ, Ridings KM, Vattiato G, et al. "Real-Time Estimation of the Effective Reproduction Number of SARS-CoV-2 in Aotearoa New Zealand." *PeerJ* 10 (October 2022):e14119.

65. Glanz J, Hvistendahl M, Chang A. "How Deadly Was China's Covid Wave?" *New York Times*. February 15, 2023. https://www.nytimes.com/interactive/2023/02/15/world/asia/china-covid-death-estimates.html

Chapter 4. The Long Road to Generative AI

1. Noy S, Zhang W. "Experimental Evidence on the Productivity Effects of Generative Artificial Intelligence." *Science* 381, no. 6654 (July 2023):187–92.

2. "Introducing Be My AI (Formerly Virtual Volunteer) for People Who Are Blind or Have Low Vision, Powered by OpenAI's GPT-4." Be My Eyes. https://www.bemyeyes.com/blog/introducing-be-my-eyes-virtual-volunteer. Accessed February 23, 2024.

3. Roose K. "A Conversation with Bing's Chatbot Left Me Deeply Unsettled." *New York Times*. February 16, 2023. https://www.nytimes.com/2023/02/16/technology/bing-chatbot-microsoft-chatgpt.html

4. "Margaret Mitchell: Google Fires AI Ethics Founder." BBC News. February 20, 2021. https://www.bbc.com/news/technology-56135817

5. Weiser B, Schweber N. "The ChatGPT Lawyer Explains Himself." *New York Times*. June 8, 2023. https://www.nytimes.com/2023/06/08/nyregion/lawyer-chatgpt-sanctions.html

6. Solaiman I, Talat Z, Agnew W, Ahmad L, Baker D, Blodgett SL, et al. "Evaluating the Social Impact of Generative AI Systems in Systems and Society." arXiv. Last revised June 12, 2023. http://arxiv.org/abs/2306.05949

7. Guingrich R, Graziano MSA. "Chatbots as Social Companions: How People Perceive Consciousness, Human Likeness, and Social Health Benefits in Machines." arXiv. Last revised December 16, 2023. http://arxiv.org/abs/2311.10599

8. Xiang C. "'He Would Still Be Here': Man Dies by Suicide after Talking with AI Chatbot, Widow Says." Vice. March 30, 2023. https://www.vice.com/en/article/pkadgm/man-dies-by-suicide-after-talking-with-ai-chatbot-widow-says

9. Buell S. "An MIT Student Asked AI to Make Her Headshot More 'Professional.' It Gave Her Lighter Skin and Blue Eyes." *Boston Globe*. July 19, 2023. https://www.bostonglobe.com/2023/07/19/business/an-mit-student-asked-ai-make-her-headshot-more-professional-it-gave-her-lighter-skin-blue-eyes/

10. Melissa Heikkilä. "The Viral AI Avatar App Lensa Undressed Me—without My Consent." *MIT Technology Review*. December 12, 2022. https://www.technologyreview.com/2022/12/12/1064751/the-viral-ai-avatar-app-lensa-undressed-me-without-my-consent/

11. Maiberg E. "Inside the AI Porn Marketplace Where Everything and Everyone Is for Sale." 404 Media. August 22, 2023. https://www.404media.co/inside-the-ai-porn-marketplace-where-everything-and-everyone-is-for-sale/

12. Allyn B. "A Robot Was Scheduled to Argue in Court. Then Came the Jail Threats." NPR. January 25, 2023. https://www.npr.org/2023/01/25/1151435033/a-robot-was-scheduled-to-argue-in-court-then-came-the-jail-threats

13. Mitchell M. *Artificial Intelligence: A Guide for Thinking Humans*. Reprint edition. New York: Picador; 2020.

14. McCulloch WS, Pitts W. "A Logical Calculus of the Ideas Immanent in Nervous Activity." *Bull Math Biophys* 5, no. 4 (December 1943):115–33.

15. Olazaran M. "A Sociological Study of the Official History of the Perceptrons Controversy." *Soc Stud Sci* 26, no. 3 (1996):611–59.

16. Samuel AL. "Some Studies in Machine Learning Using the Game of Checkers." *IBM J Res Dev* 3, no. 3 (July 1959):210–29.

17. Minsky M, Papert SA. *Perceptrons: An Introduction to Computational Geometry, Expanded Edition*. Cambridge, MA.: MIT Press; 1987.

18. Rumelhart DE, Hinton GE, Williams RJ. "Learning Representations by Back-Propagating Errors." *Nature* 323, no. 6088 (October 1986):533–36.

19. LeCun Y, Boser B, Denker JS, Henderson D, Howard RE, Hubbard W, et al. "Backpropagation Applied to Handwritten Zip Code Recognition." *Neural Comput* 1, no. 4 (December 1989):541–51.

20. Pulver D. "The Mail Must Get Through." *New Volusian*. September 13, 1992. https://news.google.com/newspapers?nid=1901&dat=19920912&id=kIgfAAAAIBAJ&pg=1970,5531361

21. Cortes C, Vapnik V. "Support-Vector Networks." *Mach Learn* 20, no. 3 (September 1995):273–97.

22. Deng J, Dong W, Socher R, Li LJ, Li K, Fei-Fei L. "Imagenet: A Large-Scale Hierarchical Image Database." In *2009 IEEE Conference on Computer Vision and Pattern Recognition*. Miami, FL: IEEE; 2009. p. 248–55. https://ieeexplore.ieee.org/document/5206848

23. Gershgorn D. "The Data That Transformed AI Research—and Possibly the World." Quartz. July 26, 2017. https://qz.com/1034972/the-data-that-changed-the-direction-of-ai-research-and-possibly-the-world

24. Russakovsky O, Deng J, Su H, Krause J, Satheesh S, Ma S, et al. "ImageNet Large Scale Visual Recognition Challenge." *Int J Comput Vis* 115, no. 3 (December 2015):211–52.

25. Gershgorn D. "The Inside Story of How AI Got Good Enough to Dominate Silicon Valley." Quartz. June 18, 2018. https://qz.com/1307091/the-inside-story-of-how-ai-got-good-enough-to-dominate-silicon-valley

26. Krizhevsky A, Sutskever I, Hinton GE. "ImageNet Classification with Deep Convolutional Neural Networks." In: *Advances in Neural Information Processing Systems*. Curran Associates; 2012. https://papers.nips.cc/paper_files/paper/2012/hash/c399862d3b9d6b76c8436e924a68c45b-Abstract.html

27. Cireşan DC, Meier U, Masci J, Gambardella LM, Schmidhuber J. "High-Performance Neural Networks for Visual Object Classification." arXiv. February 1, 2011. http://arxiv.org/abs/1102.0183

28. Hinton G, Deng L, Yu D, Dahl GE, Mohamed Abdel-rahman, Jaitly N, et al. "Deep Neural Networks for Acoustic Modeling in Speech Recognition: The Shared Views of Four Research Groups." *IEEE Signal Process Mag* 29, no. 6 (November 2012): 82–97.

29. Sun C, Shrivastava A, Singh S, Gupta A. "Revisiting Unreasonable Effectiveness of Data in Deep Learning Era." In: *2017 IEEE International Conference on Computer Vision (ICCV)*. Venice, IT: IEEE; 2017. p. 843–52. https://ieeexplore.ieee.org/document/8237359

30. "Open Repository of Web Crawl Data." Common Crawl. https://commoncrawl.org/

31. Orhon A, Wadhwa A, Kim Y, Rossi F, Jagadeesh V. "Deploying Transformers on the Apple Neural Engine." Apple Machine Learning Research. June 2022. https://machinelearning.apple.com/research/neural-engine-transformers

32. Hara K, Adams A, Milland K, Savage S, Callison-Burch C, Bigham JP. "A Data-Driven Analysis of Workers' Earnings on Amazon Mechanical Turk." In: *Proceedings of the 2018 CHI Conference on Human Factors in Computing Systems*. New York: Association for Computing Machinery; 2018. p. 1–14. https://doi.org/10.1145/3173574.3174023

33. Birhane A, Prabhu VU. Large Image Datasets: A Pyrrhic Win for Computer Vision? In: *2021 IEEE Winter Conference on Applications of Computer Vision (WACV)*. Waikoloa, HI: IEEE; 2021. p. 1536–46. https://ieeexplore.ieee.org/document/9423393

34. Yang K, Qinami K, Fei-Fei L, Deng J, Russakovsky O. "Towards Fairer Datasets: Filtering and Balancing the Distribution of the People Subtree in the ImageNet Hierarchy." In: *Proceedings of the 2020 Conference on Fairness, Accountability, and Transparency*. New York: Association for Computing Machinery; 2020. p. 547–58. https://doi.org/10.1145/3351095.3375709

35. Yang K, Yau JH, Fei-Fei L, Deng J, Russakovsky O. "A Study of Face Obfuscation in ImageNet." In: *Proceedings of the 39th International Conference on Machine Learning*. PMLR; 2022. p. 25313–30. https://proceedings.mlr.press/v162/yang22q.html

36. Grant N, Hill K. "Google's Photo App Still Can't Find Gorillas. And Neither Can Apple's." *New York Times*. May 22, 2023. https://www.nytimes.com/2023/05/22/technology/ai-photo-labels-google-apple.html

37. "Language Models and Linguistic Theories Beyond Words." *Nat Mach Intell* 5, no. 7 (July 2023):677–8.

38. Olah C, Mordvintsev A, Schubert L. "Feature Visualization." *Distill* 7, no. 2 (November 2017):10.23915/distill.00007.

39. "Stable Diffusion Launch Announcement." Stability AI. August 10, 2022. https://stability.ai/news/stable-diffusion-announcement

40. Davis W. "AI Companies Have All Kinds of Arguments against Paying for Copyrighted Content." The Verge. November 4, 2023. https://www.theverge.com/2023/11/4/23946353/generative-ai-copyright-training-data-openai-microsoft-google-meta-stabilityai

41. Nolan B. "Artists Say AI Image Generators Are Copying Their Style to Make Thousands of New Images—and It's Completely out of Their Control." *Business Insider*. October 7, 2022. https://www.businessinsider.com/ai-image-generators-artists-copying-style-thousands-images-2022-10

42. Vincent J. "Getty Images Is Suing the Creators of AI Art Tool Stable Diffusion for Scraping Its Content." The Verge. January 17, 2023. https://www.theverge.com/2023/1/17/23558516/ai-art-copyright-stable-diffusion-getty-images-lawsuit

43. Lauryn Ipsum [@LaurynIpsum]. "I'm cropping these for privacy reasons/because I'm not trying to call out any one individual. These are all Lensa portraits where the mangled remains of an artist's signature is still visible. That's the remains of the signature of one of the multiple artists it stole from. A 🏺." X (formerly Twitter). December 5, 2022. https://twitter.com/LaurynIpsum/status/1599953586699767808

44. Jiang HH, Brown L, Cheng J, Khan M, Gupta A, Workman D, et al. "AI Art and Its Impact on Artists." In: *Proceedings of the 2023 AAAI/ACM Conference on AI, Ethics, and Society*. New York: ACM; 2023. p. 363–74. https://dl.acm.org/doi/10.1145/3600211.3604681

45. Edwards B. "Artists Stage Mass Protest against AI-Generated Artwork on ArtStation." Ars Technica. December 15, 2022. https://arstechnica.com/information-technology/2022/12/artstation-artists-stage-mass-protest-against-ai-generated-artwork/

46. Fassler E. "South Korea Is Giving Millions of Photos to Facial Recognition Researchers." Vice. November 16, 2021. https://www.vice.com/en/article/xgdxqd/south-korea-is-selling-millions-of-photos-to-facial-recognition-researchers

47. Ho-sung C. S. "Korean Government Provided 170M Facial Images Obtained in Immigration Process to Private AI Developers." Hankyoreh. October 21, 2021. https://english.hani.co.kr/arti/english_edition/e_national/1016107.html

48. Inzamam Q, Qadri H. "Telangana Is Inching Closer to Becoming a Total Surveillance State." The Wire. July 15, 2022. https://thewire.in/tech/telangana-surveillance-police-cctv-facial-recognition

49. Mozur P. "One Month, 500,000 Face Scans: How China Is Using A.I. to Profile a Minority." *New York Times*. April 14, 2019. https://www.nytimes.com/2019/04/14/technology/china-surveillance-artificial-intelligence-racial-profiling.html

50. Hill K. "The Secretive Company That Might End Privacy as We Know It." *New York Times*. January 18, 2020. https://www.nytimes.com/2020/01/18/technology/clearview-privacy-facial-recognition.html

51. Hill K. *Your Face Belongs to Us: A Secretive Startup's Quest to End Privacy as We Know It*. New York: Random House; 2023.

52. Heikkilä M. "The Walls Are Closing In on Clearview AI." *MIT Technology Review*. May 24, 2022. https://www.technologyreview.com/2022/05/24/1052653/clearview-ai-data-privacy-uk/

53. Mac R. "How a Facial Recognition Tool Found Its Way into Hundreds of US Police Departments, Schools, and Taxpayer-Funded Organizations." BuzzFeed News. Updated April 9, 2021. https://www.buzzfeednews.com/article/ryanmac/clearview-ai-local-police-facial-recognition

54. DeGeurin M. "Targeted Billboard Ads Are a Privacy Nightmare." Gizmodo. October 13, 2022. https://gizmodo.com/billboards-facial-recognition-privacy-targeted-ads-1849655599

55. Big Brother Watch Team. "The Streets Are Watching: How Billboards Are Spying on You." Big Brother Watch. October 12, 2022.

56. Howard J, Ruder S. "Universal Language Model Fine-tuning for Text Classification." In: *Proceedings of the 56th Annual Meeting of the Association for Computational Linguistics (Volume 1: Long Papers)*. ed. Gurevych I, Miyao Y. Melbourne, Australia: ACL; 2018. p. 328–39. https://aclanthology.org/P18-1031

57. Raffel C, Shazeer N, Roberts A, Lee K, Narang S, Matena M, et al. "Exploring the Limits of Transfer Learning with a Unified Text-to-Text Transformer." *J Mach Learn Res* 21, no. 1 (January 2020):140:5485–140:5551.

58. Wei J, Bosma M, Zhao VY, Guu K, Yu AW, Lester B, et al. "Finetuned Language Models Are Zero-Shot Learners." arXiv; last revised February 8, 2022. http://arxiv.org/abs/2109.01652

59. Ouyang L, Wu J, Jiang X, Almeida D, Wainwright C, Mishkin P, et al. "Training Language Models to Follow Instructions with Human Feedback." *Adv Neural Inf Process Syst* 35 (December 2022):27730–44.

60. Acher M. "Debunking the Chessboard: Confronting GPTs against Chess Engines to Estimate Elo Ratings and Assess Legal Move Abilities" (blog). September 30, 2023. https://blog.mathieuacher.com/GPTsChessEloRatingLegalMoves/

61. "What Is Heavier?" r/ChatGPT. Reddit. February 26, 2023. www.reddit.com/r/ChatGPT/comments/11clqc5/what_is_heavier/

62. Jawahar G, Sagot B, Seddah D. "What Does BERT Learn about the Structure of Language?" In: *Proceedings of the 57th Annual Meeting of the Association for Computational Linguistics*, ed. Korhonen A, Traum D, Màrquez L. Florence, Italy: ACL; 2019. p. 3651–57. https://aclanthology.org/P19-1356

63. Rogers A, Kovaleva O, Rumshisky A. "A Primer in BERTology: What We Know about How BERT Works," ed. Johnson M, Roark B, Nenkova A. *Trans Assoc Comput Linguist* 8 (2020):842–66.

64. Li K, Hopkins AK, Bau D, Viégas F, Pfister H, Wattenberg M. "Emergent World Representations: Exploring a Sequence Model Trained on a Synthetic Task." arXiv. Last revised February 27, 2023. http://arxiv.org/abs/2210.13382

65. Frankfurt HG. *On Bullshit*. Princeton, NJ: Princeton University Press; 2005.

66. Verma P, Oremus W. "ChatGPT Invented a Sexual Harassment Scandal and Named a Real Law Prof as the Accused." *Washington Post*. April 14, 2023. https://www.washingtonpost.com/technology/2023/04/05/chatgpt-lies/

67. Brown EN. "The A.I. Defamation Cases Are Here: ChatGPT Sued for Spreading Misinformation." Reason. June 7, 2023. https://reason.com/2023/06/07/the-a-i-defamation-cases-are-here-chatgpt-sued-for-spreading-misinformation/

68. Bonifacic I. "CNET Corrected Most of Its AI-Written Articles." Engadget. Updated January 25, 2023. https://www.engadget.com/cnet-corrected-41-of-its-77-ai-written-articles-201519489.html

69. Christian J. "CNET Sister Site Restarts AI Articles, Immediately Publishes Idiotic Error." Futurism. Updated February 1, 2023. https://futurism.com/cnet-bankrate-restarts-ai-articles

70. Kao J. "More Than a Million Pro-Repeal Net Neutrality Comments Were Likely Faked." HackerNoon. November 22, 2017. https://hackernoon.com/more-than-a-million-pro-repeal-net-neutrality-comments-were-likely-faked-e9f0e3ed36a6

71. Weiss M. "Deepfake Bot Submissions to Federal Public Comment Websites Cannot Be Distinguished from Human Submissions." *Technol Sci* (December 2017). https://techscience.org/a/2019121801/

72. Stupp C. "Fraudsters Used AI to Mimic CEO's Voice in Unusual Cybercrime Case." *Wall Street Journal*. August 30, 2019. https://www.wsj.com/articles/fraudsters-use-ai-to-mimic-ceos-voice-in-unusual-cybercrime-case-11567157402

73. Porter J. "Apple Books Quietly Launches AI-Narrated Audiobooks." The Verge. January 5, 2023. https://www.theverge.com/2023/1/5/23540261/apple-text-to-speech-audiobooks-ebooks-artificial-intelligence-narrator-madison-jackson

74. "TikTok Voice Generator with Custom TTS Voices." Resemble AI. 2021. https://www.resemble.ai/tiktok/

75. Cox J. "AI-Generated Voice Firm Clamps down after 4chan Makes Celebrity Voices for Abuse." Vice. January 30, 2023. https://www.vice.com/en/article/dy7mww/ai-voice-firm-4chan-celebrity-voices-emma-watson-joe-rogan-elevenlabs

76. Edwards L. "Deepfakes in the Courts." Counsel. December 5, 2022. https://www.counselmagazine.co.uk/articles/deepfakes-in-the-courts

77. Ajder H, Patrini G, Cavalli F. "Automating Image Abuse: Deepfake Bots on Telegram." Sensity. October 2020.

78. Vincent J. "UK Plans to Make the Sharing of Non-consensual Deepfake Porn Illegal." The Verge. November 25, 2022. https://www.theverge.com/2022/11/25/23477548/uk-deepfake-porn-illegal-offence-online-safety-bill-proposal

79. Abid A, Farooqi M, Zou J. "Persistent Anti-Muslim Bias in Large Language Models." In: *Proceedings of the 2021 AAAI/ACM Conference on AI, Ethics, and Society.* New York: ACM; 2021. p. 298–306. https://doi.org/10.1145/3461702.3462624

80. Rowe N. "'It's Destroyed Me Completely': Kenyan Moderators Decry Toll of Training of AI Models." *Guardian* (US edition). August 2, 2023. https://www.theguardian.com/technology/2023/aug/02/ai-chatbot-training-human-toll-content-moderator-meta-openai

81. "OpenAI Software Engineer Salary | $800K-$925K+." Levels.fyi. last updated January 30, 2024. https://www.levels.fyi/companies/openai/salaries/software-engineer

82. Metz C, Mickle T. "OpenAI Completes Deal That Values the Company at $80 Billion." *New York Times.* February 16, 2024. https://www.nytimes.com/2024/02/16/technology/openai-artificial-intelligence-deal-valuation.html

83. Dzieza J. "AI Is a Lot of Work." Intelligencer. June 20, 2023. https://nymag.com/intelligencer/article/ai-artificial-intelligence-humans-technology-business-factory.html

84. Jones P. "Refugees Help Power Machine Learning Advances at Microsoft, Facebook, and Amazon." Rest of World. September 22, 2021. https://restofworld.org/2021/refugees-machine-learning-big-tech/

85. Perrigo B. "AI by the People, for the People." *Time.* July 27, 2023. https://time.com/6297403/the-workers-behind-ai-rarely-see-its-rewards-this-indian-startup-wants-to-fix-that/

86. Williams A, Miceli M, Gebru T. "The Exploited Labor behind Artificial Intelligence." Noema. October 13, 2022. https://www.noemamag.com/the-exploited-labor-behind-artificial-intelligence

87. Inbal S, GitHub staff. "Survey Reveals AI's Impact on the Developer Experience" (blog). GitHub. June 13, 2023. https://github.blog/2023-06-13-survey-reveals-ais-impact-on-the-developer-experience/

88. Narayanan A, Kapoor S, Lazar S. "Model Alignment Protects against Accidental Harms, Not Intentional Ones." AI Snake Oil. December 1, 2023. https://www.aisnakeoil.com/p/model-alignment-protects-against

Chapter 5. Is Advanced AI an Existential Threat?

1. Boak J, O'Brien M. "Biden Wants to Move Fast on AI Safeguards and Signs an Executive Order to Address His Concerns." AP News. updated October 30, 2023. https://apnews.com/article/biden-ai-artificial-intelligence-executive-order-cb86162000d894f238f28ac029005059

2. "Pause Giant AI Experiments: An Open Letter." Future of Life Institute. March 22, 2023. https://futureoflife.org/open-letter/pause-giant-ai-experiments/

3. "Statement on AI Risk." Center for AI Safety. May 30, 2023. https://www.safe.ai/statement-on-ai-risk

4. Crevier D. *AI: The Tumultuous History of the Search for Artificial Intelligence.* New York: Basic Books; 1993.

5. Marcus G. "Face It, Self-Driving Cars Still Haven't Earned Their Stripes." Marcus on AI. August 19, 2023. https://garymarcus.substack.com/p/face-it-self-driving-cars-still-havent

6. Allyn-Feuer A, Sanders T. "Transformative AGI by 2043 is <1% likely." arXiv. June 5, 2023. http://arxiv.org/abs/2306.02519

7. "How Generative Models Could Go Wrong." *Economist.* April 19, 2023. https://www.economist.com/science-and-technology/2023/04/19/how-generative-models-could-go-wrong

8. Karger E, Rosenberg J, Jacobs Z, Hickman M, Hadshar R, Gamin K, et al. "Forecasting Existential Risks Evidence from a Long-Run Forecasting Tournament." Forecasting Research Institute. July 10, 2023. https://forecastingresearch.org/news/results-from-the-2022-existential-risk-persuasion-tournament

9. Turing AM. "On Computable Numbers, with an Application to the Entscheidungsproblem." *Proc Lond Math Soc* s2–42, no. 1 (1937):230–65.

10. McCarthy J, Minsky ML, Rochester N, Shannon CE. "A Proposal for the Dartmouth Summer Research Project on Artificial Intelligence." *AI Mag* 27, no. 4 (December 2006):12–12.

11. Liu X, Yu H, Zhang H, Xu Y, Lei X, Lai H, et al. "AgentBench: Evaluating LLMs as Agents." arXiv. Last revised October 25, 2023. http://arxiv.org/abs/2308.03688

12. LeCun Y [@ylecun]. "On the highway towards Human-Level AI, Large Language Model is an off-ramp." X (formerly Twitter). February 4, 2023. https://twitter.com/ylecun/status/1621805604900585472

13. Weizenbaum J. "ELIZA—a Computer Program for the Study of Natural Language Communication between Man and Machine." *Commun ACM* 9, no. 1 (January 1966):36–45.

14. Mitchell M. "Why AI Is Harder Than We Think." arXiv. Last revised April 28, 2021. http://arxiv.org/abs/2104.12871

15. Clark J, Amodei D. "Faulty Reward Functions in the Wild." Open AI. December 21, 2016. https://openai.com/research/faulty-reward-functions

16. Christian B. *The Alignment Problem: Machine Learning and Human Values*. New York: W.W. Norton; 2020.

17. Hernandez D, Brown TB. "Measuring the Algorithmic Efficiency of Neural Networks." arXiv. May 8, 2020. http://arxiv.org/abs/2005.04305

18. "Introducing Falcon LLM." Technology Innovation Institute." 2023. https://falconllm.tii.ae/. Accessed June 1, 2023.

19. Narayanan A, Kapoor S, Lazar S. "Model Alignment Protects against Accidental Harms, Not Intentional Ones." AI Snake Oil. December 1, 2023. https://www.aisnakeoil.com/p/model-alignment-protects-against

20. Takanen A, Demott JD, Miller C. *Fuzzing for Software Security*. Norwood, MA: Artech House Publishers; 2008.

21. Root E. "The Evolution of Security: The Story of Code Red" (blog). *Kapersky Daily*, August 4, 2022. https://www.kaspersky.com/blog/history-lessons-code-red/45082/

22. Stolberg SG, Mueller B, Zimmer C. "The Origins of the Covid Pandemic: What We Know and Don't Know." *New York Times*. March 17, 2023. https://www.nytimes.com/article/covid-origin-lab-leak-china.html

Chapter 6. Why Can't AI Fix Social Media?

1. Transcript courtesy of Bloomberg Government. "Transcript of Mark Zuckerberg's Senate Hearing." *Washington Post*. April 10, 2018. https://www.washingtonpost.com/news/the-switch/wp/2018/04/10/transcript-of-mark-zuckerbergs-senate-hearing/

2. Shrivastava R. "Mastodon Isn't a Replacement for Twitter—but It Has Rewards of Its Own." *Forbes*. November 4, 2022. https://www.forbes.com/sites/rashishrivastava/2022/11/04/mastodon-isnt-a-replacement-for-twitterbut-it-has-rewards-of-its-own/

3. Masnick M. "Hey Elon: Let Me Help You Speed Run the Content Moderation Learning Curve." Techdirt. November 2, 2022. https://www.techdirt.com/2022/11/02/hey-elon-let-me-help-you-speed-run-the-content-moderation-learning-curve/

4. Gray ML, Suri S. *Ghost Work: How to Stop Silicon Valley from Building a New Global Underclass.* New York: Houghton Mifflin Harcourt; 2019.

5. Roberts ST. *Behind the Screen: Content Moderation in the Shadows of Social Media; With a New Preface.* New Haven, CT: Yale University Press; 2021.

6. Williams A, Miceli M, Gebru T. "The Exploited Labor behind Artificial Intelligence." Noema. October 13, 2022. https://www.noemamag.com/the-exploited-labor-behind-artificial-intelligence

7. Bateman J, Thompson N, Smith V. "How Social Media Platforms' Community Standards Address Influence Operations." Carnegie Endowment for International Peace. April 1, 2021. https://carnegieendowment.org/2021/04/01/how-social-media-platforms-community-standards-address-influence-operations-pub-84201

8. Newton C. "The Secret Lives of Facebook Moderators in America." The Verge. February 25, 2019. https://www.theverge.com/2019/2/25/18229714/cognizant-facebook-content-moderator-interviews-trauma-working-conditions-arizona

9. Hill K. "A Dad Took Photos of His Naked Toddler for the Doctor. Google Flagged Him as a Criminal." *New York Times.* August 21, 2022. https://www.nytimes.com/2022/08/21/technology/google-surveillance-toddler-photo.html

10. Montgomery B. "Twitter Suspends an Account for a Cartoon of Captain America Punching a Nazi." The Daily Beast. September 11, 2019. https://www.thedailybeast.com/twitter-suspends-an-account-for-tweeting-a-cartoon-of-captain-america-punching-a-nazi

11. Butler M. "Cornell Library YouTube Page Restored after Termination Last Week over Nudity Content." The Ithaca Voice. June 24, 2022. http://ithacavoice.org/2022/06/cornell-library-youtube-page-restored-after-termination-last-week-over-nudity-content/

12. Knight W. "Why a YouTube Chat about Chess Got Flagged for Hate Speech." *Wired.* March 1, 2021. https://www.wired.com/story/why-youtube-chat-chess-flagged-hate-speech/

13. Koebler J, Cox J. "The Impossible Job: Inside Facebook's Struggle to Moderate Two Billion People." Vice. August 23, 2018. https://www.vice.com/en/article/xwk9zd/how-facebook-content-moderation-works

14. The Real Facebook Oversight Board. "Content Moderators Emergency Session." YouTube video, 1:08:10. October 26, 2020. https://www.youtube.com/watch?v=F1byT_2htfs

15. Oremus W. "Facebook's Contracted Moderators Say They're Paid to Follow Orders, Not Think." OneZero. October 28, 2020. https://onezero.medium.com/facebooks-contracted-moderators-say-they-re-paid-to-follow-orders-not-think-40331991c6ee

16. Jan T, Dwoskin E. "A White Man Called Her Kids the N-word. Facebook Stopped Her from Sharing It." *Washington Post*. July 31, 2017. https://www.washingtonpost.com/business/economy/for-facebook-erasing-hate-speech-proves-a-daunting-challenge/2017/07/31/922d9bc6-6e3b-11e7-9c15-177740635e83_story.html

17. Dwoskin E, Tiku N, Timberg C. "Facebook's Race-Blind Practices around Hate Speech Came at the Expense of Black Users, New Documents Show." *Washington Post*. November 11, 2021. https://www.washingtonpost.com/technology/2021/11/21/facebook-algorithm-biased-race/

18. Horwitz J, Scheck J. "Facebook Increasingly Suppresses Political Movements It Deems Dangerous." *Wall Street Journal*. October 22, 2021. https://www.wsj.com/articles/facebook-suppresses-political-movements-patriot-party-11634937358

19. "Myanmar: The Social Atrocity: Meta and the Right to Remedy for the Rohingya." Amnesty International. September 29, 2022. https://www.amnesty.org/en/documents/asa16/5933/2022/en/

20. Allen C. "Facebook's Content Moderation Failures in Ethiopia" (blog). Council on Foreign Relations. April 19, 2022. https://www.cfr.org/blog/facebooks-content-moderation-failures-ethiopia

21. Zelalem Z, Guest P. "Why Facebook Keeps Failing in Ethiopia." Rest of World. November 13, 2021. https://restofworld.org/2021/why-facebook-keeps-failing-in-ethiopia/

22. Purnell N, Horwitz J. "Facebook Services Are Used to Spread Religious Hatred in India, Internal Documents Show." *Wall Street Journal*. October 23, 2021. https://www.wsj.com/articles/facebook-services-are-used-to-spread-religious-hatred-in-india-internal-documents-show-11635016354

23. Facebook. *Facebook Response: Sri Lanka Human Rights Impact Assessment*. May 12, 2020. https://about.fb.com/wp-content/uploads/2021/03/FB-Response-Sri-Lanka-HRIA.pdf

24. Scott M. "Facebook Did Little to Moderate Posts in the World's Most Violent Countries." POLITICO. October 25, 2021. https://www.politico.com/news/2021/10/25/facebook-moderate-posts-violent-countries-517050

25. "Bridging the Gap: Local Voices in Content Moderation." ARTICLE 19. https://www.article19.org/bridging-the-gap-local-voices-in-content-moderation/. Accessed February 17, 2024.

26. Debre I, Akram F. "Facebook's Language Gaps Weaken Screening of Hate, Terrorism." AP News. October 25, 2021. https://apnews.com/article/the-facebook-papers-language-moderation-problems-392cb2d065f81980713f37384d07e61f

27. Lorenz-Spreen P, Oswald L, Lewandowsky S, Hertwig R. "A Systematic Review of Worldwide Causal and Correlational Evidence on Digital Media and Democracy." *Nat Hum Behav* 7, no. 1 (January 2023):74–101.

28. Ortutay B. "In a 3rd Test, Facebook Still Fails to Block Hate Speech." AP News. July 28, 2022. https://apnews.com/article/technology-africa-kenya-7aaee94 59ae58e1278b6075f9ed1392b

29. Gebru T. "Moving beyond the Fairness Rhetoric in Machine Learning. Invited Talk, ICLR 2021." YouTube video, 1:10:48. September 28, 2021. https://www.youtube.com/watch?v=gfB8pOZkFLE

30. Santa Clara Principles. "Santa Clara Principles on Transparency and Accountability in Content Moderation." 2021. https://santaclaraprinciples.org/. Accessed February 18, 2024.

31. Tworek H. "History Explains Why Global Content Moderation Cannot Work." Brookings. December 10, 2021. https://www.brookings.edu/articles/history-explains-why-global-content-moderation-cannot-work/

32. "Understanding When Content Is Withheld Based on Country." Help Center. Twitter. https://help.twitter.com/en/rules-and-policies/post-withheld-by-country. Accessed September 28, 2023.

33. Sakunia S. "Twitter Blocked 122 Accounts in India at the Government's Request." Rest of World. March 24, 2023. https://restofworld.org/2023/twitter-blocked-access-punjab-amritpal-singh-sandhu/

34. Fisher M. "Inside Facebook's Secret Rulebook for Global Political Speech." *New York Times*. December 27, 2018. https://www.nytimes.com/2018/12/27/world/facebook-moderators.html

35. Sumbaly R, Miller M, Shah H, Xie Y, Culatana SC, Khatkevich T, et al. "Using AI to Detect COVID-19 Misinformation and Exploitative Content" (blog). ML Applications. May 12, 2020. https://ai.meta.com/blog/using-ai-to-detect-covid-19-misinformation-and-exploitative-content/

36. Richmond R. "Web Gang Operating in the Open." *New York Times*. January 16, 2012.

37. Eady G, Paskhalis T, Zilinsky J, Bonneau R, Nagler J, Tucker JA. "Exposure to the Russian Internet Research Agency Foreign Influence Campaign on Twitter in the 2016 US Election and Its Relationship to Attitudes and Voting Behavior." *Nat Commun* 14, no. 1 (January 2023):62.

38. Gerrard Y. "Beyond the Hashtag: Circumventing Content Moderation on Social Media." *New Media Soc* 20, no. 12 (December 2018):4492–511.

39. Franklin JC, Ribeiro JD, Fox KR, Bentley KH, Kleiman EM, Huang X, et al. "Risk Factors for Suicidal Thoughts and Behaviors: A Meta-Analysis of 50 Years of Research." *Psychol Bull* 143, no. 2 (February 2017):187–232.

40. Cummings J. "Prevent Suicide by Recognizing Early Warning Signs." Cummings Institute. 2016. https://cgi.edu/biodyne-model-therapists-masters-suicide-assessment-prevention/

41. Nyren E. "Pete Davidson Posts Unsettling Message, Deletes Instagram." *Variety*. December 15, 2018. https://variety.com/2018/tv/news/pete-davidson-deletes-instagram-1203090685/

42. Bryant M. "What Are Social Media Companies Doing about Suicidal Posts?" *Guardian* (US edition). January 10, 2019. https://www.theguardian.com/society/2019/jan/10/cupcakke-hospitalised-twitter-social-media-suicidal-posts

43. Card C. "How Facebook AI Helps Suicide Prevention." Meta. September 10, 2018. https://about.fb.com/news/2018/09/inside-feed-suicide-prevention-and-ai/

44. Kaste M. "Facebook Increasingly Reliant on A.I. to Predict Suicide Risk." NPR. November 17, 2018. https://www.npr.org/2018/11/17/668408122/facebook-increasingly-reliant-on-a-i-to-predict-suicide-risk

45. Fuller DA, Lamb HR, Biasotti M, Snook J. "Overlooked in the Undercounted: The Role of Mental Illness in Fatal Law Enforcement Encounters." Treatment Advocacy Center. December 2015. https://www.treatmentadvocacycenter.org/reports_publications/overlooked-in-the-undercounted-the-role-of-mental-illness-in-fatal-law-enforcement-encounters/

46. Marks M. "Artificial Intelligence-Based Suicide Prediction." *Yale J Law and Tech* 21, no. 3 (2019):98–121.

47. Olofsson B, Jacobsson L. "A Plea for Respect: Involuntarily Hospitalized Psychiatric Patients' Narratives about Being Subjected to Coercion." *J Psychiatr Ment Health Nurs* 8, no. 4 (August 2001):357–66.

48. "Section 230." Electronic Frontier Foundation. https://www.eff.org/issues/cda230. Accessed February 17, 2024.

49. "Viacom Sues Google, YouTube for $1 Billion." NBC News. March 13, 2007. https://www.nbcnews.com/id/wbna17592285

50. Lee TB. "How YouTube Lets Content Companies 'Claim' NASA Mars Videos." Ars Technica. August 8, 2012. https://arstechnica.com/tech-policy/2012/08/how-youtube-lets-content-companies-claim-nasa-mars-videos/

51. Brodeur MA. "Copyright Bots and Classical Musicians Are Fighting Online. The Bots Are Winning." *Washington Post*. May 21, 2020. https://www.washingtonpost.com/entertainment/music/copyright-bots-and-classical-musicians-are-fighting-online-the-bots-are-winning/2020/05/20/a11e349c-98ae-11ea-89fd-28fb313d1886_story.html

52. Glaze V. "MrBeast Calls Out YouTube after Being Hit with False Copyright Strike." Dexerto. February 12, 2019. https://www.dexerto.com/entertainment/mrbeast-calls-out-youtube-after-being-hit-false-copyright-strike-357775/

53. WatchMojo.com. "Are Rights Holders Unlawfully Claiming Billions in AdSense Revenue?" YouTube video, 28:09. May 9, 2019. https://www.youtube.com/watch?v=-w1f3olwqcg

54. Wodinsky S. "YouTube's Copyright Strikes Have Become a Tool for Extortion." The Verge. February 11, 2019. https://www.theverge.com/2019/2/11/18220032/youtube-copystrike-blackmail-three-strikes-copyright-violation

55. Cushing T. "Cops Are Still Playing Copyrighted Music to Thwart Citizens Recording Their Actions." Techdirt. April 18, 2022. https://www.techdirt.com/2022/04/18/cops-are-still-playing-copyrighted-music-to-thwart-citizens-recording-their-actions/

56. Gillespie T. *Custodians of the Internet: Platforms, Content Moderation, and the Hidden Decisions That Shape Social Media*. New Haven, CT: Yale University Press; 2018.

57. Gillespie T. *Custodians of the Internet: Platforms, Content Moderation, and the Hidden Decisions That Shape Social Media*. New Haven, CT: Yale University Press; 2018.

58. Klonick K. "The Facebook Oversight Board: Creating an Independent Institution to Adjudicate Online Free Expression." *Yale Law J* 129, no. 8 (June 2020):2418–99. https://www.yalelawjournal.org/feature/the-facebook-oversight-board

59. adam22 [@adam22]. "How to get your Instagram back if it gets deleted." X (formerly Twitter.) May 18, 2022. https://twitter.com/adam22/status/1527005564802600960

60. nadalizadeh. "Google has terminated our Developer Account, says it is 'associated'?" r/androiddev. Reddit post. March 30, 2022. www.reddit.com/r/androiddev/comments/ts6jfg/google_has_terminated_our_developer_account_says/

61. "Why We Are Striking." Etsy Strike. July 14, 2022 (capture date: archived). https://web.archive.org/web/20220714164329/https://etsystrike.org/join-us/why-we-are-striking/

62. Facebook. *Facebook's Response to the Oversight Board's First Decisions*. February 2021. https://about.fb.com/wp-content/uploads/2021/02/OB_First-Decision_Detailed_.pdf

63. Elliott V. "Big Tech Ditched Trust and Safety. Now Startups Are Selling It Back as a Service." *Wired*. November 6, 2023. https://www.wired.com/story/trust-and-safety-startups-big-tech/

64. Papada E, Altman D, Angiolillo F, Gastaldi L, Köhler T, Lundstedt M, et al. "Defiance in the Face of Autocratization." (V-Dem Institute Working Paper—Democracy Report, University of Gothenburg, Gothenburg, Sweden, 2023). https://papers.ssrn.com/abstract=4560857

65. "The World's Most, and Least, Democratic Countries in 2022." *Economist*. February 1, 2023. https://www.economist.com/graphic-detail/2023/02/01/the-worlds-most-and-least-democratic-countries-in-2022

66. Menn J, Shih G. "Under India's Pressure, Facebook Let Propaganda and Hate Speech Thrive." *Washington Post*. September 26, 2023. https://www.washingtonpost.com/world/2023/09/26/india-facebook-propaganda-hate-speech/

67. McCordick J. "Twitter's Elon Musk Defends Decision to Limit Tweets in Turkey during Tight Presidential Election." *Vanity Fair*. May 14, 2023. https://www.vanityfair.com/news/2023/05/twitter-musk-censors-turkey-election-erdogan

68. Keller D. "Six Things about Jawboning." (blog). Knight First Amendment Institute at Columbia University. October 10, 2023. http://knightcolumbia.org/blog/six-things-about-jawboning

69. Tollefson J. "Disinformation Researchers under Investigation: What's Happening and Why." *Nature*. July 5, 2023. https://www.nature.com/articles/d41586-023-02195-3

70. Narayanan A. "Understanding Social Media Recommendation Algorithms." Knight First Amendment Institute at Columbia University. March 9, 2023. http://knightcolumbia.org/content/understanding-social-media-recommendation-algorithms

71. Ovadya A, Thorburn L. "Bridging Systems: Open Problems for Countering Destructive Divisiveness across Ranking, Recommenders, and Governance." Knight First Amendment Institute at Columbia University. October 26, 2023. http://knightcolumbia.org/content/bridging-systems

72. Malik A. "Twitter Begins Rolling Out Its Community Notes Feature Globally." TechCrunch. December 12, 2022. https://techcrunch.com/2022/12/12/twitter-begins-rolling-out-its-community-notes-feature-globally/

73. "Community Notes: A Collaborative Way to Add Helpful Context to Posts and Keep People Better Informed." X (formerly Twitter). https://communitynotes.twitter.com/guide/en/about/introduction. Accessed February 17, 2024.

74. Milli S, Carroll M, Wang Y, Pandey S, Zhao S, Dragan AD. "Engagement, User Satisfaction, and the Amplification of Divisive Content on Social Media." arXiv. Last revised December 22, 2023. http://arxiv.org/abs/2305.16941

75. Zuckerberg M. "A Blueprint for Content Governance and Enforcement." Facebook. Last edited May 5, 2021. https://www.facebook.com/notes/751449002072082/

76. Iyer R. "Content Moderation Is a Dead End." Designing Tomorrow. October 7, 2022. https://psychoftech.substack.com/p/content-moderation-is-a-dead-end

77. Stray J, Iyer R, Larrauri HP. "The Algorithmic Management of Polarization and Violence on Social Media." Knight First Amendment Institute at Columbia University. August 22, 2023. http://knightcolumbia.org/content/the-algorithmic-management-of-polarization-and-violence-on-social-media

78. Joshi AR. "Overrun by Influencers, Historic Sites Are Banning TikTok Creators in Nepal." Rest of World. July 18, 2022. https://restofworld.org/2022/nepals-historic-sites-banning-tiktok-creators/

79. Keller D. "The DSA's Industrial Model for Content Moderation." Verfassungsblog. February 24, 2022. https://verfassungsblog.de/dsa-industrial-model/

80. Douek E. "Content Moderation as Systems Thinking." *Harvard Law Review*. December 2022. https://harvardlawreview.org/print/vol-136/content-moderation-as-systems-thinking/

81. Matias JN. "The Civic Labor of Volunteer Moderators Online." *Soc Media Soc* 5, no. 2 (April 2019):2056305119836778.

82. Witynski M. "Unpaid Social Media Moderators Perform Labor Worth at Least $3.4 Million a Year on Reddit Alone." Northwestern Now. May 31, 2022. https://news.northwestern.edu/stories/2022/05/unpaid-social-media-moderators/

Chapter 7. Why Do Myths about AI Persist?

1. Rudd KE, Johnson SC, Agesa KM, Shackelford KA, Tsoi D, Kievlan DR, et al. "Global, Regional, and National Sepsis Incidence and Mortality, 1990–2017: Analysis for the Global Burden of Disease Study." *Lancet* 395, no: 10219 (January 2020): 200–11.

2. "Unbundling Epic: How the Electronic Health Record Market Is Being Disrupted." Research. CB Insights. August 4, 2021. https://www.cbinsights.com/research/report/electronic-health-record-companies-unbundling/

3. Williams N. "Health Tech Giant Epic Systems Is Focusing on Machine Learning. Here's Why." *Milwaukee Business Journal*. August 5, 2019. https://www.bizjournals.com/milwaukee/news/2019/08/05/health-tech-giant-epic-systems-is-focusing-on.html

4. Cleveland Clinic. "Virtual Ideas for Tomorrow | Judy Faulkner, CEO and Founder, EPIC." YouTube video, 33:20. September 2, 2020. https://www.youtube.com/watch?v=BXnw15pGv-U

5. Wong A, Otles E, Donnelly JP, Krumm A, McCullough J, DeTroyer-Cooley O, et al. "External Validation of a Widely Implemented Proprietary Sepsis Prediction Model in Hospitalized Patients." *JAMA Intern Med* 181, no. 8 (August 2021): 1065–70.

6. Gerhart J, Thayer J. "For Clinicians, by Clinicians: Our Take on Predictive Models." Epic. June 28, 2021. https://www.epic.com/epic/post/for-clinicians-by-clinicians-our-take-on-predictive-models

7. Drees J. "Epic Pays Hospitals That Use Its EHR Algorithms, Report Finds." Becker's Health IT. July 26, 2021. https://www.beckershospitalreview.com/ehrs/epic-pays-hospitals-that-use-its-ehr-algorithms-report-finds.html

8. Ross C. "Epic's AI Algorithms, Shielded from Scrutiny by a Corporate Firewall, Are Delivering Inaccurate Information on Seriously Ill Patients." STAT. July 26, 2021. https://www.statnews.com/2021/07/26/epic-hospital-algorithms-sepsis-investigation/

9. Ross C. "Epic Overhauls Popular Sepsis Algorithm Criticized for Faulty Alarms." STAT. October 3, 2022. https://www.statnews.com/2022/10/03/epic-sepsis-algorithm-revamp-training/

10. Murray SG, Wachter RM, Cucina RJ. "Discrimination by Artificial Intelligence in a Commercial Electronic Health Record—a Case Study." Health Aff Forefr. January 31, 2020. https://www.healthaffairs.org/do/10.1377/forefront.20200128.626576/full/

11. Solon O. "The Rise of 'Pseudo-AI': How Tech Firms Quietly Use Humans to Do Bots' Work." *Guardian* (US edition). July 6, 2018. https://www.theguardian.com/technology/2018/jul/06/artificial-intelligence-ai-humans-bots-tech-companies

12. "How It Works." x.ai. May 18, 2021 (archived). https://archive.is/jkKBI

13. Huet E. "The Humans Hiding behind the Chatbots." Bloomberg. April 18, 2016. https://www.bloomberg.com/news/articles/2016-04-18/the-humans-hiding-behind-the-chatbots

14. Johnson K. "Government Audit of AI with Ties to White Supremacy Finds No AI." VentureBeat. April 5, 2021. https://venturebeat.com/business/government-audit-of-ai-with-ties-to-white-supremacy-finds-no-ai/

15. Ryan M. "In AI We Trust: Ethics, Artificial Intelligence, and Reliability." *Sci Eng Ethics* 26, no. 5 (October 2020):2749–67.

16. Rozenblit L, Keil F. "The Misunderstood Limits of Folk Science: An Illusion of Explanatory Depth." *Cogn Sci* 26, no. 5 (September 2002):521–62.

17. "Gartner Hype Cycle Research Methodology." Gartner. https://www.gartner.com/en/research/methodologies/gartner-hype-cycle. Accessed February 23, 2024.

18. Mullany M. "8 Lessons from 20 Years of Hype Cycles." LinkedIn. December 7, 2016. https://www.linkedin.com/pulse/8-lessons-from-20-years-hype-cycles-michael-mullany/

19. Huang J, O'Neill C, Tabuchi H. "Bitcoin Uses More Electricity Than Many Countries. How Is That Possible?" *New York Times*. September 3, 2021. https://www.nytimes.com/interactive/2021/09/03/climate/bitcoin-carbon-footprint-electricity.html

20. Spangler T. "Larry David, Tom Brady, Stephen Curry, Other Celebs Sued over FTX Crypto Exchange Collapse." *Variety*. November 16, 2022. https://variety.com/2022/digital/news/ftx-lawsuit-larry-david-tom-brady-stephen-curry-crypto-1235434627/

21. Kaloudis G. "I'm Glad There Are No Crypto Super Bowl Ads: Here's Why." CoinDesk. February 12, 2023. https://www.coindesk.com/consensus-magazine/2023/02/12/im-glad-there-are-no-crypto-super-bowl-ads-heres-why/

22. White M. "Web3 Is Going Just Great." https://www.web3isgoinggreat.com/. Accessed February 22, 2024.

23. James Lighthill. *Part I Artificial Intelligence: A General Survey*. July 1972. https://www.aiai.ed.ac.uk/events/lighthill1973/lighthill.pdf

24. Mitchell M. "Why AI is Harder than We Think." arXiv. Last revised April 28, 2021. http://arxiv.org/abs/2104.12871

25. Whittaker M. "The Steep Cost of Capture." *Interactions* 28, no. 6 (November 2021):50–55.

26. Laufer B, Jain S, Cooper AF, Kleinberg J, Heidari H. "Four Years of FAccT: A Reflexive, Mixed-Methods Analysis of Research Contributions, Shortcomings, and Future Prospects." In: *2022 ACM Conference on Fairness, Accountability, and Transparency*. Seoul Republic of Korea: ACM; 2022. p. 401–26. https://dl.acm.org/doi/10.1145/3531146.3533107

27. Ahmed N, Wahed M, Thompson NC. "The Growing Influence of Industry in AI Research." *Science* 379, no. 6635 (March 2023):884–86.

28. Myers BA. "A Brief History of Human-Computer Interaction Technology." *Interactions* 5, no. 2 (March 1998):44–54.

29. Lundh A, Lexchin J, Mintzes B, Schroll JB, Bero L. "Industry Sponsorship and Research Outcome." *Cochrane Database Syst Rev* 2, no. 2 (February 2017): MR000033.

30. Dickinson J. "Deadly Medicines and Organised Crime." *Can Fam Physician* 60, no. 4 (April 2014):367–68.

31. Sismondo S. *Ghost-Managed Medicine*. Manchester, UK: Mattering Press; 2018. https://www.matteringpress.org/books/ghost-managed-medicine

32. Rahimi A, Recht B. "Reflections on Random Kitchen Sinks." *arg min blog*. December 5, 2017. http://archives.argmin.net/2017/12/05/kitchen-sinks/

33. Lipton ZC, Steinhardt J. "Troubling Trends in Machine Learning Scholarship: Some ML Papers Suffer from Flaws That Could Mislead the Public and Stymie Future Research." *Queue* 17, no. 1 (February 2019):80:45–80:77.

34. Creative Destruction Lab. "Geoff Hinton: On Radiology." YouTube video, 1:24. November 24, 2016. https://www.youtube.com/watch?v=2HMPRXstSvQ

35. Henderson M. "Radiology Facing a Global Shortage." RSNA News. May 10, 2022. https://www.rsna.org/news/2022/may/Global-Radiologist-Shortage

36. Franta B. "Shell and Exxon's Secret 1980s Climate Change Warnings. *Guardian* (US edition). September 19, 2018. https://www.theguardian.com/environment/climate-consensus-97-per-cent/2018/sep/19/shell-and-exxons-secret-1980s-climate-change-warnings

37. Hiltzik M. "A New Study Shows How Exxon Mobil Downplayed Climate Change When It Knew the Problem Was Real." *Los Angeles Times*. August 22, 2017. https://www.latimes.com/business/hiltzik/la-fi-hiltzik-exxonmobil-20170822-story.html

38. InfluenceMap. *An Investor Enquiry: How Much Big Oil Spends on Climate Lobbying.* April 2016. https://influencemap.org/report/Climate-Lobbying-by-the-Fossil-Fuel-Sector

39. Winecoff AA, Watkins EA. "Artificial Concepts of Artificial Intelligence: Institutional Compliance and Resistance in AI Startups." In: *Proceedings of the 2022 AAAI/ACM Conference on AI, Ethics, and Society.* New York: ACM; 2022. p. 788–99. https://dl.acm.org/doi/10.1145/3514094.3534138

40. Winecoff AA, Watkins EA. "Artificial Concepts of Artificial Intelligence: Institutional Compliance and Resistance in AI Startups." In: *Proceedings of the 2022 AAAI/ACM Conference on AI, Ethics, and Society.* New York: ACM; 2022. p. 788–99. https://dl.acm.org/doi/10.1145/3514094.3534138. Ellipses in original.

41. Grill G. "Constructing Certainty in Machine Learning: On the Performativity of Testing and Its Hold on the Future." OSF Preprints; created September 7, 2022. https://osf.io/zekqv/

42. Raji D, Denton E, Bender EM, Hanna A, Paullada A. "AI and the Everything in the Whole Wide World Benchmark." *Proc Neural Inf Process Syst Track Datasets Benchmarks* 1 (December 2021). https://datasets-benchmarks-proceedings.neurips.cc/paper/2021/hash/084b6fbb10729ed4da8c3d3f5a3ae7c9-Abstract-round2.html

43. OpenAI. "GPT-4 Technical Report." arXiv. Last revised December 19, 2023. http://arxiv.org/abs/2303.08774

44. Bratman B. "Improving the Performance of the Performance Test: The Key to Meaningful Bar Exam Reform." *UMKC Law Review* 83 (April 2015):565. https://papers.ssrn.com/abstract=2520042

45. Camerer CF, Dreber A, Forsell E, Ho TH, Huber J, Johannesson M, et al. "Evaluating Replicability of Laboratory Experiments in Economics." *Science* 351, no. 6280 (March 2016):1433–6.

46. Camerer CF, Dreber A, Holzmeister F, Ho TH, Huber J, Johannesson M, et al. "Evaluating the Replicability of Social Science Experiments in *Nature* and *Science* Between 2010 and 2015." *Nat Hum Behav* 2, no. 9 (September 2018):637–44.

47. Muchlinski D, Siroky D, He J, Kocher M. "Comparing Random Forest with Logistic Regression for Predicting Class-Imbalanced Civil War Onset Data." *Polit Anal* 24, no. 1 (2016):87–103.

48. Colaresi M, Mahmood Z. "Do the Robot: Lessons from Machine Learning to Improve Conflict Forecasting." *J Peace Res* 54, no. 2 (March 2017):193–214.

49. Kaufman AR, Kraft P, Sen M. "Improving Supreme Court Forecasting Using Boosted Decision Trees." *Polit Anal* 27, no. 3 (July 2019):381–7.

50. Kapoor S, Narayanan A. "Leakage and the Reproducibility Crisis in Machine-Learning-Based Science." *Patterns* 4, no. 9 (September 2023):100804.

51. Kapoor S, Nanayakkara P, Peng K, Pham H, Narayanan A. "The Reproducibility Crisis in ML-based Science." Workshop with slides, Princeton University, Princeton, NJ, July 28, 2022. https://sites.google.com/princeton.edu/rep-workshop

52. Kapoor S, Narayanan A. "OpenAI's Policies Hinder Reproducible Research on Language Models." AI Snake Oil. March 22, 2023. https://www.aisnakeoil.com/p/openais-policies-hinder-reproducible

53. Kapoor S, Cantrell E, Peng K, Pham TH, Bail CA, Gundersen OE, et al. "REFORMS: Reporting Standards for Machine Learning Based Science." arXiv. Last revised September 19, 2023. http://arxiv.org/abs/2308.07832

54. Kapoor S, Narayanan A. "Eighteen Pitfalls to Beware of in AI Journalism." AI Snake Oil. September 30, 2022. https://www.aisnakeoil.com/p/eighteen-pitfalls-to-beware-of-in

55. Willyard C. "Can AI Fix Electronic Medical Records?" *Scientific American*. February 1, 2020. https://www.scientificamerican.com/article/can-ai-fix-electronic-medical-records/

56. Murphy K. "Epic's Faulkner Has High Hopes for Forthcoming Cosmos Technology." TechTarget. September 3, 2020 (archived). https://archive.is/3f72G

57. Bender EM. "On NYT Magazine on AI: Resist the Urge to be Impressed." Medium. April 17, 2022. https://medium.com/@emilymenonbender/on-nyt-magazine-on-ai-resist-the-urge-to-be-impressed-3d92fd9a0edd

58. Smith CS. "A.I. Here, There, Everywhere." *New York Times*. February 23, 2021. https://www.nytimes.com/2021/02/23/technology/ai-innovation-privacy-seniors-education.html

59. York C. "Algorithm Claims to Predict Crime in US Cities Before It Happens." Bloomberg. June 30, 2022. https://www.bloomberg.com/news/articles/2022-06-30/new-algorithm-can-predict-crime-in-us-cities-a-week-before-it-happens

60. Mirror Now Digital. "Minority Report Soon? New AI Tech to Predict Crimes Weeks ahead with 90% Accuracy." Times Now. July 1, 2022. https://www.timesnownews.com/mirror-now/in-focus/minority-report-soon-new-ai-tech-to-predict-crimes-weeks-ahead-with-90-accuracy-article-92599864

61. Hogg R. "AI Model Predicting Crime in US Cities Is Right Nine Times out of 10." Business Insider. July 3, 2022. https://www.businessinsider.com/ai-model-predicts-crime-us-nine-times-out-of-ten-2022-7

62. Thubron R. "Newly Developed Algorithm Able to Predict Crime a Week in Advance with 90% Accuracy." Techspot. July 1, 2022. https://www.techspot.com/community/topics/newly-developed-algorithm-able-to-predict-crime-a-week-in-advance-with-90-accuracy.276016/

63. Wood M. "Algorithm Predicts Crime a Week in Advance, but Reveals Bias in Police Response." UChicago. June 30, 2022. https://biologicalsciences.uchicago.edu/news/algorithm-predicts-crime-police-bias

64. Sumner P, Vivian-Griffiths S, Boivin J, Williams A, Venetis CA, Davies A, et al. "The Association between Exaggeration in Health Related Science News and Academic Press Releases: Retrospective Observational Study. *BMJ* 349 (December 10, 2014):g7015.

65. Woolston C. "Study Points to Press Releases as Sources of Hype." *Nature* 516, no. 7531 (December 2014):291–291.

66. Ardia D, Ringel E, Ekstrand VS, Fox A. "Addressing the Decline of Local News, Rise of Platforms, and Spread of Mis- and Disinformation Online." The Center for Information, Technology, and Public Life (CITAP). https://citap.unc.edu/news/local-news-platforms-mis-disinformation/. Accessed December 22, 2020.

67. Kissinger H, Schmidt E, Huttenlocher DP, Schouten S. *The Age of AI: And Our Human Future*. New York: Little Brown and Company; 2021.

68. Whittaker M, Suchman L. "The Myth of Artificial Intelligence." The American Prospect. December 8, 2021. https://prospect.org/api/content/7fc7f7c2-5781-11ec-987e-12f1225286c6/

69. Vinsel L. "You're Doing It Wrong: Notes on Criticism and Technology Hype." Medium. February 1, 2021. https://sts-news.medium.com/youre-doing-it-wrong-notes-on-criticism-and-technology-hype-18b08b4307e5

70. "Pause Giant AI Experiments: An Open Letter." Future of Life Institute. March 22, 2023. https://futureoflife.org/open-letter/pause-giant-ai-experiments/

71. Kennedy B, Tyson A, Saks E. "Public Awareness of Artificial Intelligence in Everyday Activities." Pew Research Center. February 15, 2023. https://www.pewresearch.org/science/2023/02/15/public-awareness-of-artificial-intelligence-in-everyday-activities/

72. Jakesch M, Hancock JT, Naaman M. "Human Heuristics for AI-Generated Language Are Flawed." *Proc Natl Acad Sci* 120, no. 11 (March 2023):e2208839120.

73. Sellier AL, Scopelliti I, Morewedge CK. "Debiasing Training Improves Decision Making in the Field." *Psychol Sci* 30, no. 9 (September 2019):1371–9.

74. Morewedge CK, Yoon H, Scopelliti I, Symborski CW, Korris JH, Kassam KS. "Debiasing Decisions: Improved Decision Making with a Single Training Intervention." *Policy Insights Behav Brain Sci* 2, no. 1 (October 2015):129–40.

Chapter 8. Where Do We Go from Here?

1. Tarnoff B. *Internet for the People: The Fight for Our Digital Future*. London: Verso; 2022.

2. Yin L, Sankin A. "Dollars to Megabits, You May Be Paying 400 Times as Much as Your Neighbor for Internet Service." The Markup. October 19, 2022. https://themarkup.org/still-loading/2022/10/19/dollars-to-megabits-you-may-be-paying-400-times-as-much-as-your-neighbor-for-internet-service

3. "Community Network Map." Community Networks. Muni numbers updated September 2021. https://communitynets.org/content/community-network-map

4. Rajendra-Nicolucci C, Sugarman M, Zuckerman E. "The Three-Legged Stool: A Manifesto for a Smaller, Denser Internet." Initiative for Digital Public Infrastructure. March 29, 2023. https://publicinfrastructure.org/2023/03/29/the-three-legged-stool/

5. Broussard M. *Artificial Unintelligence: How Computers Misunderstand the World.* First MIT Press paperback edition. Cambridge, MA: The MIT Press; 2019.

6. " Newspapers Fact Sheet." Pew Research Center. November 10, 2023. https://www.pewresearch.org/journalism/fact-sheet/newspapers/

7. García E, Kraft MA, Schwartz HL "Are We at a Crisis Point with the Public Teacher Workforce? Education Scholars Share Their Perspectives." Brookings. August 26, 2022. https://www.brookings.edu/articles/are-we-at-a-crisis-point-with-the-public-teacher-workforce-education-scholars-share-their-perspectives/

8. Liang W, Yuksekgonul M, Mao Y, Wu E, Zou J. "GPT Detectors Are Biased against Non-native English Writers." *Patterns* 4, no. 7 (July 2023). https://www.cell.com/patterns/abstract/S2666-3899(23)00130-7

9. Jiminez K. "Professors Are Using ChatGPT Detector Tools to Accuse Students of Cheating. But What If the Software Is Wrong?" *USA Today*. April 12, 2023. https://www.usatoday.com/story/news/education/2023/04/12/how-ai-detection-tool-spawned-false-cheating-case-uc-davis/11600777002/

10. Verma P. "A Professor Accused His Class of Using ChatGPT, Putting Diplomas in Jeopardy." *Washington Post*. May 19, 2023. https://www.washingtonpost.com/technology/2023/05/18/texas-professor-threatened-fail-class-chatgpt-cheating/

11. Gramlich J. "What the Data Says about Gun Deaths in the U.S." Pew Research Center. 2023. https://www.pewresearch.org/short-reads/2023/04/26/what-the-data-says-about-gun-deaths-in-the-u-s/

12. Gee G. "AI Tries (and Fails) to Detect Weapons in Schools." The Intercept. May 7, 2023. https://theintercept.com/2023/05/07/ai-gun-weapons-detection-schools-evolv/

13. Bani A. "Philadelphia is Allocating Hundreds of Millions of Dollars to Address Mounting Gun Violence." The Plug. December 30, 2022. https://tpinsights.com/philadelphia-is-allocating-hundreds-of-millions-of-dollars-to-address-mounting-gun-violence/

14. "ShotSpotter Frequently Asked Questions." SoundThinking (formerly ShotSpotter). January 2018. https://www.soundthinking.com/faqs/shotspotter-faqs/

15. Cushing T. "Chicago PD Oversight Says ShotSpotter Tech Is Mostly Useless When It Comes to Fighting Gun Crime." Techdirt. August 26, 2021. https://www

.techdirt.com/2021/08/26/chicago-pd-oversight-says-shotspotter-tech-is-mostly-useless-when-it-comes-to-fighting-gun-crime/

16. Dodge J, De Mar C, Hickey M. "Mayor Johnson Cancels Controversial Chicago Gunshot Detection System." CBS News. February 13, 2024. https://www.cbsnews.com/chicago/news/mayor-johnson-cancels-controversial-chicago-gunshot-detection-system/

17. Wootson Jr. CR. "Charlotte Ends Contract with ShotSpotter Gunshot Detection System." *Charlotte Observer*. February 10, 2016. https://www.charlotteobserver.com/news/local/crime/article59685506.html

18. Davila V. "S.A. Cuts Funding to $550K Gunshot Detection Program That Resulted in 4 Arrests." MySA. August 15, 2017. https://www.mysanantonio.com/news/local/article/City-pulls-plug-on-pricey-gunshot-detection-system-11817475.php

19. Kalven J. "Chicago Awaits Video of Police Killing of 13-Year-Old Boy." The Intercept. April 13, 2021. https://theintercept.com/2021/04/13/chicago-police-killing-boy-adam-toledo-shotspotter/

20. Burke G, Mendoza M, Linderman J, Tarm M. "How AI-Powered Tech Landed Man in Jail with Scant Evidence." AP News. March 5, 2022. https://apnews.com/article/artificial-intelligence-algorithm-technology-police-crime-7e3345485aa668c97606d4b54f9b6220

21. Cheves H. "ShotSpotter Is a Failure. What's Next?" MacArthur Justice Center. May 5, 2022. https://www.macarthurjustice.org/blog2/shotspotter-is-a-failure-whats-next/

22. Sen A, Bennett DK. "The Black Box: Colleges Spend Thousands on AI to Prevent Suicides and Shootings. Evidence That It Works Is Scant." *Dallas Morning News*. December 1, 2022. https://interactives.dallasnews.com/2022/the-black-box-social-sentinel/

23. Irani L, Alexander K. "The Oversight Bloc." *Logic(s) Magazine*. December 25, 2021. https://logicmag.io/beacons/the-oversight-bloc/

24. Reich R, Sahami M, Weinstein JM. *System Error: Where Big Tech Went Wrong and How We Can Reboot*. New York: Harper; 2021.

25. Schwartz B. "Top Colleges Should Select Randomly from a Pool of 'Good Enough.'" *Chronicle of Higher Education*. February 25, 2005. https://bschwartz.domains.swarthmore.edu/Chronicle%20of%20Higher%20Education%202-25-05.pdf

26. Gross K, Bergstrom CT. "Contest Models Highlight Inherent Inefficiencies of Scientific Funding Competitions." *PLOS Biol* 17, no. 1 (January 2019):e3000065.

27. Baicker K, Taubman SL, Allen HL, Bernstein M, Gruber JH, Newhouse JP, et al. "The Oregon Experiment—Effects of Medicaid on Clinical Outcomes." *N Engl J Med* 368, no. 18 (May 2013):1713–22.

28. Henderson H. "Why Cash Payments Aren't Always the Best Tool to Help Poor People." The Conversation. March 17, 2021. http://theconversation.com/why-cash-payments-arent-always-the-best-tool-to-help-poor-people-156019

29. Uzogara EE. "Democracy Intercepted." *Science* 381, no. 6656 (July 2023): 386–7.

30. Meinhardt C, Lawrence CM, Gailmard LA, Zhang D, Bommasani R, Kosoglu R, et al. "By the Numbers: Tracking the AI Executive Order." HAI. November 16, 2023. https://hai.stanford.edu/news/numbers-tracking-ai-executive-order

31. Sheehan M. "China's AI Regulations and How They Get Made." Carnegie Endowment for International Peace. July 10, 2023. https://carnegieendowment.org/2023/07/10/china-s-ai-regulations-and-how-they-get-made-pub-90117

32. "FTC Action Stops Business Opportunity Scheme That Promised Its AI-Boosted Tools Would Power High Earnings through Online Stores." Press release. Federal Trade Commission. August 22, 2023. https://www.ftc.gov/news-events/news/press-releases/2023/08/ftc-action-stops-business-opportunity-scheme-promised-its-ai-boosted-tools-would-power-high-earnings

33. Lomas N. "FTC Settlement with Ever Orders Data and AIs Deleted after Facial Recognition Pivot." TechCrunch. January 12, 2021. https://techcrunch.com/2021/01/12/ftc-settlement-with-ever-orders-data-and-ais-deleted-after-facial-recognition-pivot/

34. Oremus W. "OpenAI CEO Tells Senate That He Fears AI's Potential to Manipulate Views." *Washington Post*. May 16, 2023. https://www.washingtonpost.com/technology/2023/05/16/ai-congressional-hearing-chatgpt-sam-altman/

35. Bickert M. "Charting a Way Forward on Online Content Regulation." Meta. February 17, 2020. https://about.fb.com/news/2020/02/online-content-regulation/

36. Constine J. "Facebook Asks for a Moat of Regulations It Already Meets." TechCrunch. February 17, 2020. https://techcrunch.com/2020/02/17/regulate-facebook/

37. Keck K. "Big Tobacco: A History of Its Decline." CNN. June 19, 2009. https://edition.cnn.com/2009/POLITICS/06/19/tobacco.decline/

38. Brownell KD, Warner KE. "The Perils of Ignoring History: Big Tobacco Played Dirty and Millions Died. How Similar Is Big Food?" *Milbank Q* 87, no. 1 (March 2009):259–94.

39. McKinnon JD, Day C. "Tech Companies Make Final Push to Head Off Tougher Regulation." *Wall Street Journal*. December 19, 2022 https://www.wsj.com/articles/tech-companies-make-final-push-to-head-off-tougher-regulation-11671401283

40. Evans B. "AI and the Automation of Work." Benedict Evans. July 2, 2023. https://www.ben-evans.com/benedictevans/2023/7/2/working-with-ai

41. Aratani L. "US Eating Disorder Helpline Takes Down AI Chatbot over Harmful Advice." *Guardian* (US edition). May 31, 2023. https://www.theguardian.com/technology/2023/may/31/eating-disorder-hotline-union-ai-chatbot-harm

42. Verma P, Vynck GD. "ChatGPT Took Their Jobs. Now They Walk Dogs and Fix Air Conditioners." *Washington Post*. June 5, 2023. https://www.washingtonpost.com/technology/2023/06/02/ai-taking-jobs/

43. Sorgi G, Sario FD. "Who Killed the EU's Translators?" POLITICO. May 12, 2023. https://www.politico.eu/article/translators-translation-european-union-eu-autmation-machine-learning-ai-artificial-intelligence-translators-jobs/

44. Bessen J. "How Computer Automation Affects Occupations: Technology, Jobs, and Skills." CEPR. September 22, 2016. https://cepr.org/voxeu/columns/how-computer-automation-affects-occupations-technology-jobs-and-skills

45. Gray ML, Suri S. *Ghost Work: How to Stop Silicon Valley from Building a New Global Underclass*. Boston: Houghton Mifflin Harcourt; 2019.

46. Wheeler G. "'Autocomplete on Steroids': Ted Chiang Addresses Phenomenon of AI at Granfalloon Festival." *Indiana Daily Student*. June 9, 2023. https://www.idsnews.com/article/2023/06/buskirk-chumley-theater-event-ted-chiang-talk-2023-granfalloon

47. Hong S-ha. "Prediction as Extraction of Discretion." *Big Data Soc* 10, no. 1 (January 2023):20539517231171053.

48. Maddaus G. "SAG-AFTRA Approves Deal to End Historic Strike." *Variety*. November 8, 2023. https://variety.com/2023/biz/news/sag-aftra-tentative-deal-historic-strike-1235771894/

49. Jarvey N, Press J. "Labor Pains and Gains: The Winners and Losers of the Hollywood Strikes." *Vanity Fair*. November 16, 2023. https://www.vanityfair.com/hollywood/2023/10/writers-strike-winners-and-losers

50. Allas T, Maksimainen J, Manyika J, Singh N. "An Experiment to Inform Universal Basic Income." McKinsey & Company. September 15, 2020. https://www.mckinsey.com/industries/social-sector/our-insights/an-experiment-to-inform-universal-basic-income

51. Dube A. "A Plan to Reform the Unemployment Insurance System in the United States." Brookings. April 12, 2021. https://www.brookings.edu/articles/a-plan-to-reform-the-unemployment-insurance-system-in-the-united-states/

52. Porter E. "Don't Fight the Robots. Tax Them." *New York Times*. February 23, 2019. https://www.nytimes.com/2019/02/23/sunday-review/tax-artificial-intelligence.html

53. Acemoglu D, Johnson S. "Big Tech Is Bad. Big A.I. Will Be Worse." *New York Times*. June 9, 2023. https://www.nytimes.com/2023/06/09/opinion/ai-big-tech-microsoft-google-duopoly.html

54. Lohr S. "Economists Pin More Blame on Tech for Rising Inequality." *New York Times*. January 11, 2022. https://www.nytimes.com/2022/01/11/technology/income-inequality-technology.html

55. McQuillan D. *Resisting AI: An Anti-fascist Approach to Artificial Intelligence*. Bristol, UK: Bristol University Press; 2022.

56. Boyd D, Hargittai E, Schultz J, Palfrey J. "Why Parents Help Their Children Lie to Facebook about Age: Unintended Consequences of the 'Children's Online Privacy Protection Act.'" First Monday. October 31, 2011. https://firstmonday.org/ojs/index.php/fm/article/view/3850

57. Boyd D. "Protect Elders! Ban Television!!" Medium. April 10, 2023. https://zephoria.medium.com/protect-elders-ban-television-2b18ab49988b

58. Thompson B. "Instagram, TikTok, and the Three Trends." Stratechery. August 16, 2022. https://stratechery.com/2022/instagram-tiktok-and-the-three-trends/

59. Twenge JM, Farley E. "Not All Screen Time Is Created Equal: Associations with Mental Health Vary by Activity and Gender." *Soc Psychiatry Psychiatr Epidemiol* 56, no. 2 (February 2021):207–17.

60. Mollick ER, Mollick L. "Assigning AI: Seven Approaches for Students, with Prompts." SSRN. September 23, 2023. https://papers.ssrn.com/abstract=4475995

61. Watters A. "The 100 Worst Ed-Tech Debacles of the Decade." Hack Education. December 31, 2019. http://hackeducation.com/2019/12/31/what-a-shitshow

62. Keener G. "Chess Is Booming." *New York Times*. June 17, 2022. https://www.nytimes.com/2022/06/17/crosswords/chess/chess-is-booming.html

63. Ortiz K. *Written Testimony of Karla Ortiz: US. Senate Judiciary Subcommittee on Intellectual Property "AI and Copyright."* July 7, 2023. https://www.judiciary.senate.gov/imo/media/doc/2023-07-12_pm_-_testimony_-_ortiz.pdf

64. Singer N. "Chatbot Hype or Harm? Teens Push to Broaden A.I. Literacy." *New York Times*. December 13, 2023. https://www.nytimes.com/2023/12/13/technology/ai-chatbots-schools-students.html

Epilogue to the Paperback Edition

1. Moreau T, Sinatra R, Sekara V. "Failing Our Youngest: On the Biases, Pitfalls, and Risks in a Decision Support Algorithm Used for Child Protection." In: *2024 ACM Conference on Fairness, Accountability, and Transparency*. Rio de Janeiro Brazil: ACM; 2024.

2. Closson T. "Nevada Asked A.I. Which Students Need Help. The Answer Caused an Outcry." *New York Times*. October 11, 2024.

3. Abbott J. "When Students Get Lost in the Algorithm: The Problems with Nevada's AI School Funding Experiment" (blog post). *New America*. April 3, 2024.

https://www.newamerica.org/education-policy/edcentral/when-students-get-lost-in-the-algorithm-the-problems-with-nevadas-ai-school-funding-experiment/

4. Murgia M. "Algorithms Are Deciding Who Gets Organ Transplants. Are Their Decisions Fair?" *Financial Times Magazine*. November 9, 2023.

Attia A, et al. "Effect of Recipient Age on Prioritisation for Liver Transplantation in the UK: A Population-Based Modelling Study." *The Lancet Healthy Longevity* 5, no. 5 (2024): e346–e355.

5. Narayanan A, Ströbl B, Kapoor S. "Is AI Progress Slowing Down?" AI Snake Oil newsletter. December 18, 2024.

6. Ritchie H. "What's the Impact of Artificial Intelligence on Energy Demand?" Sustainability by Numbers Newsletter. November 18, 2024.

7. Kapoor S, Narayanan A. "We Looked at 78 Election Deepfakes. Political Misinformation Is Not an AI Problem" (blog post). Knight First Amendment Institute at Columbia University. December 13, 2024. https://knightcolumbia.org/blog/we-looked-at-78-election-deepfakes-political-misinformation-is-not-an-ai-problem

8. "Experts: Watch out for AI-Generated Fakes and Disinformation About Voting Ahead of Election Day." News Literacy Project. October 29, 2024. https://newslit.org/newsroom/press-release/the-news-literacy-project-experts-watch-out-for-ai-generated-fakes-and-disinformation-about-voting-ahead-of-election-day/

9. Amodei D. "Machines of Loving Grace: How AI Could Transform the World for the Better." October 2024. https://darioamodei.com/machines-of-loving-grace

INDEX

Page numbers in *italics* refer to figures and tables.

access journalism, 25
account suspension, unaccountable, 213–14
accuracy: news media, misleading the public, 248–50; predictions, accuracy of, 67–69; transparency, 240–41
advanced AI, existential threat: accelerating progress, 165–68; AI safety meme, *170*; concluding thoughts, 177–78; expert's opinions, 151–56; global ban on, 172–74; introduction, 150–51; ladder of generality, 156–62, *160*, 162–65, *163*; notable historical computers, 157–59; reframe the issue from intelligence to power, *171*; responses to, 172–74; rogue AI, 168–71; specific threats, defending against, 174–77. *See also* Artificial General Intelligence (AGI)
Affordable Care Act (2010), 53
Afghanistan, hate speech (Facebook), 190
Age of AI, The (Huttenlocher, Kissinger, Schmidt), 251–53

agency, 69–70, 247, 253, 279
AI: agents, 163–64, 169; for content moderation, 181; content moderation, seven shortcomings for, 216–18; directions for change, 33–34; evading content moderation, 198–201; for image classification, 75, 127–29; journalism, 251; safety community, 156, 170, 172, 173; safety meme, *170*; seven reasons why content moderation is hard, *217*; for surveillance, 127–29; for translation, 191
AI, Algorithmic, and Automation Incidents and Controversies Repository, 38
AI, future of: AI snake oil, appealing to broken institutions, 261–65; embracing randomness, 265–68; generative AI, 258–60; Kai's world, 281–84; Maya's world, 285–89; overview of, 33–34; predictive AI, 261
AI and the future of work: automation, 277–78; automation paradox, 277; cloud computing, 276; copywriters

AI and the future of work (*continued*) and translators, 277; generative AI, 276–77; Hollywood actors and writers, 279; National Eating Disorders Association, 277; robot tax, 280; scriptwriting, 279; unions and workers collectives, 279; Universal Basic Income, 279–80

AI community, culture and history of hype: AI research, corporate funding of, 236–37; perceptron, 235; scientific understanding, lack of, 237–39; springs (peaks), 235; winters (valleys), 235

AI hype: AI community, culture and history of hype, 235–39; AI hypes and harms, illustrative applications, 29; cognitive biases, 255–57; Gartner hype cycle, 232; hype cycle, 229–30; hype vortex, 21–26; media, 25–26; previous technology hype, different from, 231–35; public figures spreading, 251–55. *See also* AI myths, why do they persist

AI Incident Database, 38

AI myths, why do they persist: AI community, culture and history of hype, 235–39; AI hype, 231–35; Bing chat, misleading news headlines, 249; cognitive biases, 255–57; Gartner hype cycle, 232; introduction, 227–31; news media, 247–51; public figures, 251–55; reproducibility crisis, 241–47; robots featured in news media, 248; transparency, lack of incentives for, 239–41

AI research: corporate funding, 236–37; recurring problems, 238; reproducibility crisis, 241–47

AI snake oil: 1905 advertisement for, 27; appealing to broken institutions, 261–65; definition of, 2, 26–34; embracing randomness, 265–68; hype and harms, 29

AI Snake Oil: advanced AI, existential threat of, 150–78; conclusion, 289–90; the future, predicting, 60–98; generative AI, 99–149; going forward, 258–90; introduction, 1–35; myths about AI, 227–57; predictive AI, 36–59; social media, 179–226

AI-generated: QR code image, *100*; text, 262–63; videos, 142–43; voices, 142

AISnakeOil.com, 26, 290

Alexander, Khalid, 265

AlexNet, 113

Algorithmic, and Automation Incidents and Controversies Repository, 38

algorithms: advanced AI, 161–62; crime prediction algorithm, 249–50; definition of, 39; generative AI, 109–10, 114, 116, 117, 133; predicting the future, 60–61; predictive AI, 38–40, 48–49; social media, 207, 219, 220–21

algospeak, 201

alignment, 172–73, 174–75

Allegheny County (Pennsylvania), 52–53

Allegheny Family Screening Tool, 52–53, 55

Allstate, use of predictive AI, 10–11

Altman, Sam, 274

Amazon, AI-generated books, 6

Amazon Mechanical Turk, 111, 115

American Edge, 275

Amnesty International, 189

anchoring bias, 256–57

Anderson, Wes, 7

Animal Farm (Orwell), 82

annotators, 115, 144–45
Anthropic, 137, 151, 260, 269
Apple, 116, 286
Artificial General Intelligence (AGI):
AI safety community, 156; cognitive bias, 153; definition of, 150–51; edge cases, 154; forecasting tournament, 154–55; ladder of generality, 160; risk, estimating probability of, 155; selection bias, 152–53
Artificial Intelligence Act (AIA), 271
artificial intelligence (AI), introduction to: AI hype vortex, 21–26; ChatGPT, 3–6; dawn of, 3–6; definition of, 26–34; definition of, humorous, 14; entertainment, AI shakes up, 7–9; examples of, 14; facial recognition, 15–17; generative AI, 3–6; hype and harms, illustrative applications, 29; introduction, 1–3; labeling, 12–17; labor displacement, 14; predictive AI, 2, 3, 9–12; schools and colleges, 8–9; series of curious circumstances, 18–21; snake oil advertisement (1905), 27; targeted readers (AI Snake Oil), 34–35
artificial superintelligence, 151
ArtStation, 126, 126
arXiv, 162–63
Asimov, Isaac, *Foundation*, 90, 91
Associated Press, 210, 264
AT&T, 259
ATMs, 277
attorneys, AI use, 102
autocomplete, 29, 132
autogenerated language, 140
automated decision-making, 28–29
automated hiring tools, 18, 24, 46–47
automatic translation, 191
automating bullshit: autogenerated language, 140–41; bullshit, definition of, 139; ChatGPT, 139, 140; CNET, 140; defamation by chatbot, 140; examples of, 139–41
automating script writing, 8
automation: AI and the future of work, 277–78; automation bias, 255; content moderation and, 181; hate speech, directed at Black people, 186; hiring automation, 18, 24, 66; labor concerns, 276–77, 278; low-wage workers, effect on, 280; positive effects, 277–78; robot tax, 280
automation bias, 50–51, 255
automation paradox, 277
automation's last mile, 278
autonomous driving, 13, 14, 153–54
Axios, 21

backpropagation, 109, 113
Balenciaga Pope, 7
Bankman-Fried, Sam, 234
Bard (renamed Gemini), 5, 135, 136–37
Barocas, Solon, 24, 42–43
base model, 133–34
basic reproduction number (COVID), 94–95, 96
Be My Eyes (app), 99–100
benchmark datasets, 112, 237–38, 241
benchmarking, 112, 116
Bender, Emily, 248
Bennett, Derêka K., 26
Better Business Bureau, 213
biases: AI lack of, 79; benchmarks, 116; COMPAS, 79, 80; criminal risk prediction tools, 11; image generators, 103; of policing, 80; selection bias, 152. *See also* cognitive biases
Biden, Joe, 150
Bing's chatbot, 25, 101, 249
biological risk, 176–77
Bitcoin, 233–34

Bitcoin mining, 234
Black, Rebecca, Friday (song), 88
blackout challenge, 219
blasphemy, 192–93
Bledsoe, Drew, 82
Bloomberg, 249
Bosnia, content moderation failures, 190
bots: AI agents, 163–64; benchmarks, 241; companion bots, 102; generative AI, 101; rock-paper-scissors, 137; scraping, 115. *See also* chatbots
Brady, Tom, 82
broken institutions: AI-generated text, 262; definition of, 263; educational institutions, 262–63; efficiency as a selling point, 263; gun violence and, 263; journalism, 262; law enforcement, 263–64; predictive AI, 265; snake oil, demand for, 261–62; state of hiring, 261–62; weapons, using AI for detecting, 263
Broward County (COMPAS), 79–80
brute-force intervention, 74, 131
bug finding tools, 175–76
bullshit: as autogenerated language, 140–41; ChatGPT, 139, 140; CNET, 140; defamation by chatbot, 140; definition of, 139; examples of, 139–41; generators of, 197
Butterfly Effect, 63

Canada, 128–29, 280
cancer, 69, 239
Center for AI Safety, 152
Centers for Disease Control and Prevention (CDC), 92, 93, 215
Chai, 102–3
chance events, 75
channels, automated notifications of copyright claims, 209

Charlie Bit My Finger, 86–87
Charter, 259
chatbots: arXiv, 163; factual information weakness, 5; inappropriate uses, 6; introduction, 2; National Eating Disorders Association, 277
chatbots, generative AI: Bard (renamed Gemini), 137; base model, 133–34; bullshit, automating, 139–41; Claude, 137; defamation by chatbot, 140; fine tuning, 134–35, 137–38; inaccurate outputs, 147–48; inappropriate outputs, 101–2; internal representations, 138–39; limitation of, 136–37; a meta task, fine-tuned for, 135–36; Othello games, 138; philosophical dimension, 137; practical dimension, 137; translation tool, 134–35
ChatGPT: autocomplete, 132; automating bullshit, 139, 140; Be My Eyes, 100; computer programmers, 4–5; educator's curricula, 262–63; fine tuning, 135–36; G stands for generative, 133; humor, detecting, 185; introduction, 3–6; ladder of generality, at time of writing, *163*; misinformation, 102; P stands for pretrained, 135; schools and colleges, 9; T stands for Transformer, 131; text generation, 131; token, generation of, 133
Chattanooga, Tennessee, 260
checkers-playing program, 107
chess-playing computer, 108
Chiang, Ted, 278
Chicago (ShotSpotter), 264
Chicago Tribune, 1948 election, 56
child sexual abuse, 194–95
child sexual abuse imagery, 194

Children's Online Privacy Protection Act (COPPA), 282
China: AI, oversight of, 271; chatbots, use of, 271; Core Socialist Values, 271; COVID deaths, unpredictability of, 95–97; Cybersecurity Law, 271; facial recognition, 127; misinformation, 193; social media apps banned, 216
civil wars, 23, 189, 243–44, 250–51
classifier: child sex abuse, 194–95; COVID vaccines, 196; image recognition classifier, 173; leakage example, Russian and American tanks, 244; movie review sentiment classifier, 134; slur words, discerning, 186; suicide or self-injury, imminent, 203, 204; text generation, 129
Claude (chatbot), rock-paper-scissors, 137
Clean Air Act, 269
Clean Water Act, 269
Clearview AI, 127, 128–29
clickbait, 248, 249
cliodynamics, 91–92
cloud computing, 276
CNET, 140, 262
CNN, 247
cobra population, reducing, 46
Code Red, 176
Codex, 245
cognitive bias, 153, 255–57
cognitive biases: anchoring bias, 256–57; conclusion, 257; explanatory depth, illusion of, 255–56; Future of Life Institute, 256; halo effect, 256; humans, 231; illusory truth effect, 256; overview of, 231, 255; priming, 256; quantification bias, 257
Comcast, 259

commercial AI snake oil, 23
Common Crawl, 114
community networks, 259–60
Community Notes, 220
companion chatbots, 102–3
COMPAS, 41, 55, 79–80, 239
computational predictions, effectiveness of, 67
computer vision, 111–13
computers, notable historical, 157–59
computers, predicting the future with, 62–67
Consumer Finance Protection Bureau, 270
Content ID: abuse by police officers, 209; algorithm, 207; fair use concept, 208; overview of, 206–7; running amok, 207–8
content moderation: amplification of problematic content (Zuckerberg), 221; classifiers, 184–85; a dead end, 223; evading, 198–201; future of, 223–26; hate speech, 186, 222; mistakes, 183–84; problem of their own making (companies), 218–23; process of, 181–83, 182; seven shortcomings of AI, 216–18, 217; social media's problem, 218–23; stages of AI-based content moderation cycle, 214; subreddits, 224–25, 226
content moderators, 182, 182, 185, 190–91, 214–15, 223
context, discerning, 183–88
Cook County (Illinois), 52
Correctional Offender Management Profiling for Alternative Sanctions (COMPAS), 41, 55, 79–80, 239
counter notice, 208
COVID interventions, case study (New Zealand), 96

COVID pandemic prediction: basic reproduction number, 94–95; COVID interventions, case study (New Zealand), 96; deaths, unpredictability of, 95–97; flu, difference between, 94; overview of, 93; short-term forecasting, 93–94; strategic decisions, 98

COVID-19: detector, 22–23; lab-leak theory, 177; machine learning, 22–23; misinformation, 195–96

creative labor, generative AI appropriates, 122–26

criminal justice, 40–41, 78–81, 266

criminal risk assessment, 68–69

criminal risk prediction, 11, 51–53, 59, 66, 256

criminal risk prediction systems: Allegheny Family Screening Tool, 52–53, 55; Ohio Risk Assessment System, 51; Public Safety Assessment, 51, 52

criminal risk prediction tools, 11

criti-hype, 178, 253, 254

cryptocurrency, 233–35

cultural: competence, 192; incompetence, 188–94; products, 83–84, 98

culture: AI Images, 126; equitable distribution of consumption, 83; and history of hype, 235–39; ImageNet, 114–18; information security, 176; memes, 87; Reddit, 225

cumulative advantage, 83–84, 85. *See also* rich-get-richer dynamics

Custodians of the Internet (Gillespie), 210

cybersecurity, 175–76, 177

Dallas Morning News, 26

Dall-E, 2, 7, 121, 123

D'Amelio, Charli, 88

Damon, Matt, 234

DARPA, 259

data annotation, 111, 145, 146

data centers, surveilling, 173

data collection: Google, 20; governments, 77; NSA, 78; predictive AI companies, 45; tech companies, 77

data leakage, 22

datasets: AI research, 167; benchmark datasets, 112, 237, 241; ImageNet, 114; ImageNet competition, 121–22; predictive AI, 41–42; scraped datasets, downside to, 115–16, 122; social datasets, 76; sociological datasets, 71; Stable Diffusion, 122

David, Larry, 234

deceptive practices, 25, 273

deep learning: algorithms, 117; computer vision, 113; image classification, 119; ImageNet Challenge, 74, 112–13; ladder of generality, 161–62, *162*, *163*; radiologists, replacing, 238–39; text generation, 129. *See also* ImageNet, technical and cultural significance of

deep neural networks, 109, 114, 118, *120*

deep synthesis systems, 271

deepfakes, fraud, malicious uses: AI-generated videos, 143; AI-generated voices, 142; deepfakes, 142–43, 148; scams, 142

defense in depth, 176

Deng, Jia, 111

Department of Commerce, 270

Department of Homeland Security, 17, 270

Diary of a Young Girl, The (Frank), 82

diffusion model, 121, *122*

digit recognition, 109, 110

digit recognition machines, 109

digital infrastructure, 259, 260

Digital Markets Act (DMA), 270–71
Digital Services Act of 2023 (DSA), 223–24, 270
disease prediction, 92
Disney, 83
dog whistles, 191
DoorDash, 213
Douek, Evelyn, 224
downranking or demotion, 183

EAB Navigate, 37, 38
earthquakes, 68
edge cases, 154
EdTech, 286
eight billion problem, 76, 97
election forecasting, 56
ELIZA, 165–66
ELIZA effect, 166
emotion recognition, 16
entertainment, 7–9, 88, 89, 148, 226
Epic, 227, 239
Epic sepsis model, 227–29, 230
Ethiopia, 189, 191, 260
Etsy, 213
EU, 260, 270–71
Evans, Benedict, 276
Executive Office of the President, 270
existential risks, 31, 151, 154, 255. *See also* advanced AI, existential threat
expert systems, 161, 236

Facebook: American Edge, 275; Black people, getting locked out, 187; community standards document, 181, 195, *196*; content moderation failures, 188; content moderation rules, 185–86; content moderators, 190–91, 191–92; hate speech, 189, 190, 218; Koobface, 199; Middle East, takedowns of terrorist content, 190; oversight board, 212; Phan Thi Kim Phuc (Napalm Girl), 210, 211; post removal, *182*, 182–83; regulatory capture, 274; Rohingya (Myanmar), persecution of, 189; role in society, 179; Sri Lanka, hate speech, 190; suicide prevention, 202–3; suppression of conservative political movements, 187; Tigray civil war, role in, 189; welfare checks, 203, 204; worst of the worst investigation, 187
facial analysis technology, physical billboards, 128
facial recognition: accuracy of, 15–16; banning, 129; benefits of, 17; China, 127; Clearview AI, 127, 128; danger of, 16; false arrests, 15; introduction, 15–17; police officers abuse of, 127–28; racial bias, 15; Rite Aid, falsely accusing customers, 17; surveillance, abusive use of, 127–29
fair use provision (U.S. copyright law), 123, 208
false dichotomy, cutting through, 268–74
Faulkner, Judith, 228
Federal Trade Commission (FTC), 17, 25, 140
feedback loops, 97–98, 108–9, 268
Ferguson, Viana, 186
FICO (Fair, Issac, and Company), 66
FICO score, 66
fine-tuning, 118, 134, 135–36, 137–38, 147, 173
fingerprint matching, 194, 206, 218
Finland (UBI), 280
Firefly (Adobe), 7
First Amendment (United States Constitution), 193, 272
flawed democracy (India), 216

Flu Trends, 20
FluSight competition, 92, 93
Food and Drug Administration (FDA), 28, 270
Forrester, Jay, 64
Foundation (Asimov), 90
4chan, 142, 180
Fragile Families Challenge, 70–73, 97
Fragile Families Challenge, disappointment in the end, 73–78
Frank, Anne, *The Diary of a Young Girl*, 82
Frankfurt, Harry, 139
Friday (Black), 88
FTX, 234
the future, getting specific: cancer, 69; criminal risk assessment, 68–69; earthquakes, 68; genetic diseases, 69; life outcomes, 69–70; people's futures, 67; phenomena can predict, 67; phenomena can't predict, 67; predictions, judging accuracy of, 67–69; the weather, 67–68
Future of Life Institute, 151, 254, 256

gaming, 47–48
gaming, AI incentivizes, 46–48
Gangnam Style (PSY), 88
Gartner hype cycle, 231–33, *232*
gatekeepers, 88–89
Gebru, Timnit, 102
Gemini, 5, 135, 136–37
gender identification, 16
General Data Protection Regulation (GDPR), 270
Generation Alpha, 282
generation time, 95
generative AI: AI-generated image, QR code, *100*; appropriating creative labor, 122–26, *124–25, 126*; automating bullshit, 139–41; cost of improvement, 143–46; deepfakes, fraud, malicious uses, 142–43; failure and revival, 107–10, *110*; functional QR code, *100*; historical background, 105–7, *106*; image classifying and generating, 118–22, *119, 120, 121, 122*; from images to text, 129–33; introduction, 2; in large-scale labor, 143–46; models, 123, 143, 260; from models to chatbots, 133–39; overview of, 3–6, 30, 99–105; for surveillance, 127–29, 147; taking stock, 146–49; text-to-image generation, 7–8; training machines to see, 111–13
genetic diseases, predicting, 69
Getty Images, 123
Ghost Work (Gray, Suri), 278
Gig City, 260
Gillespie, Tarleton, *Custodians of the Internet*, 210–11
GitHub Copilot, 5
Google: account suspension, unaccountable, 213–14; chatbot (Bard), 5, 137; child sexual abuse, classifier for, 194–95; dataset sizes, 114; internal criticism, silencing, 102; Mark and his toddler, 183–84, 195, 212; public knowledge to trade secrets, shift from, 260; Viacom lawsuit, copyright violation, 206; voice assistant, 248
Google Cloud, content moderation mistake, 183–84
Google Drive, 276
Google Flu Trends, 20
Google Photos, 115–16
Google Translate, 191
GPT-3, 144

GPT-4: alignment, negating the effect of, 175; human-level performance, 241; hype by public figures, 254; limitation of, 136; professional exams, 276
gradient descent, 117, 162
graphics processors (GPUs), 113, 131
Gray, Mary L., *Ghost Work*, 278
Grisham, John, 82
Gundersen, Odd Erik, 243

halo effect, 256
Harry Potter (Rowling), 82
hate speech: Afghanistan, 190; content moderation, 186, 222; cost of improvement, 143; Facebook, 189, 190, 218; inciting violence, 222; social media platforms, 181; Sri Lanka, 190; Tigray civil war, 189
health information, misinformation, 197
healthcare decisions, 43–45
herding effect, 161
Herzegovina, content moderation failures, 190
Hill, Kashmir, *Your Face Belongs to Us*, 16
Hiltzik, Michael, 247
Hinton, Geoffrey, 109, 112–13, 238–39
HireVue, 24–25, 42, 239, 261–62
hiring automation, 18, 24, 66
Hitler, A., *Mein Kampf*, 142
Holocaust denial, 192
human: content moderators, 182, 185, 214–15, 223; ingenuity, 198–201; oversight, 50
human extinction. *See* Advanced AI, existential threat
Huttenlocher, Daniel, *The Age of AI*, 251
hype, 235–39

Idaho (autogenerated language), 141
illusory truth effect, 256
image classification, 118–20, 127–29
image generation, 7, 12–13, 119, 121–22, 123, 124–25
image generators: biases and stereotypes, 103; Dall-E, 2, 121–22; downside of, 103; fake movie trailers, 7; introduction, 7–9; Midjourney, 2; Stable Diffusion, 2, 121–22
ImageNet, 111, 161
ImageNet, technical and cultural significance of, 114–18
ImageNet Challenge, 74
ImageNet competition, 111–13, 118, 121–22
images, generative AI: AI-generated image, QR code, *100*, 100–101; Be My Eyes, 100; ChatGPT, 100; classifying and generating, 118–22; Clearview AI, 127; copied without compensation, 123, 126; Dall-E 2, 123; deepfakes, 142–43; fair use provision, 122–23; generative AI models, 123; images to text, 129–33; Lensa, 103; perceptron, 106; pornographic, 103; Stability AI, 122; Stable Diffusion, 122, 123; watermarked, 123; web scraping, 115
images, introduction to, 7, 22–23
images, myths: leakage, 244; perceptron, 235; robots, 247, *248*, 256
images, social media: child sexual abuse imagery, 194–95; content moderation, 180; context, taken out of, 183–84; Facebook's internal documents, 185–86; Napalm Girl, 209–11
images to text: autocomplete, 132–33; deep learning, 129; introduction, 129–30; long-range dependencies, 130–31; processing matrices, 131

incentive misalignment, 12
incitement to violence, 187, 195
India, 46, 127, 145, 190, 216
individuals to aggregates: cliodynamics, 91–92; demand forecasting, 90; disease prediction, 92; geopolitical events, predictions about, 91; limits to prediction, reasons for, 97–98; pandemic prediction (COVID), 93–97; psychohistory, 90
Indonesia, content moderation failures, 190
information security, 175, 176
inner monologue, 163
Instagram, 203, 213, 222
Institute of Advanced Study, 63
instruction following, 135–36, 144, 172–73
internet connectivity, 259
interventions: alleviating poverty, 70; COVID interventions, case study (New Zealand), 96; lotteries, use of, 268; misinformation interventions, 197; schools, 37; suicide prevention, 201–6; welfare checks, 204–5
iPad kids, 282
iPhone 13 Pro, 115
Iraq, cultural incompetence, 190
irreducible error, 69, 75, 76
Iyer, Ravi, 222, 223

James, William, 238
Jelinek, Frederick, 116
job-candidate-assessment model, 239
John Carter, 83

Kai's world, growing up with AI, 281–84
Kapoor, Sayash, 19–21, 188
Karya, 145

Keller, Daphne, 224
Kenya, content moderation failures, 190
kidney transplant matching, 11–12
killer robots, 247, 256
Kim Phuc, Phan Thi, 209
Kindel, Alex, 72–73
Kissinger, Henry, *The Age of AI*, 251
Kjensmo, Sigbjørn, 243
Koobface, 199
Krizhevsky, Alex, 112–13

labels, 15–16, 119, 132, 194–95
labor displacement, 14, 277, 279
labor exploitation, 143–46, 148
labor rights, 35, 269
ladder of generality: first rungs, *160*; notable historical computers, *157–59*; overview of, 156–62; at time of writing, *163*; until early 2010s, *162*; what's next, 162–65
Landi, Nicolette, 16–17
language models, 141, 143–44
lawyers, 16–17, 212, 241, 276
leakage, 22, 244
LeCun, Yann, 153, 164
legacy admission, 287–88
Lensa, 103
Li, Fei-Fei, 111
liar's dividend, 148
life and death, a matter of, 201–5
life outcomes, predictability of, 75
life outcomes, predicting, 69–70
Lighthill, James, 235
Limits to Prediction, 19
Lipton, Zachary, Troubling Trends in Machine Learning Scholarship, 238
Live Time, 230
long-range dependencies, 130–31
long-term predictive power, 73–78
Lorenz, Edward, 63

low-wage workers, 180, 276, 278, 280
Lucas, George, 82–83
luck, 81–82, 85, 86, 137, 267
Lundberg, Ian, 72–73

machine learning: algorithms, 39; child sexual abuse, 194–95; COVID-19, 22–23; criminal risk prediction, 66; definition of, 13; divorce, predicting, 70–71; the future, predicting, 60–61, 65–66; hate speech, 195; hiring automation, 66; hypertension, predicting, 44–45; intervention, 24; ladder of generality, 160; music industry, predicting hit songs, 21–22; personalized ads, serving, 66; random guessing, 22; suicide prevention, 202–3
machine-learning-based: research, 22; science workshop, 23–24
machines, 109, 111–13, 277
Madison Square Garden (Abuse of AI), 16–17
main character of Twitter, 89
Marcus, Gary, 154
Mark and his toddler, 183–84, 195, 212
Markup, The, 259
Maryland, use of predictive AI, 10–11
mass collaboration, 72
Mastodon, 179, 225–26, 260
Maya's world, growing up with AI, 285–89
McCulloch, Warren, 105
McLanahan, Sara, 71–73
media: AI hype, 25–26, 231; AI hype vortex, 21, 23, 25; Flu Trends, 20; killer robots, 256; state-sponsored influence operations, 199. *See also* social media
media influencers, 88

Medicare, 9–10, 268
Mein Kampf (Hitler), 142
meme lottery, the, 86–90
meritocratic society, 83
Meta, 203, 269
Michigan, overautomation failure, 49
Microsoft, 5, 143, 176
Middle East, takedowns of terrorist content, 190
Midjourney, 2, 7
Minority Report, 75, 128
Minsky, Marvin, 153, 235
misinformation: AI hype vortex, 21–26; ChatGPT, 102; illusory truth effect, 256. *See also* automating bullshit
misinformation, social media: authoritarian governments use of, 197; COVID-19, 195–96; cultural incompetence, 193; detection of, 197; health information, 197; language models, 197; machine learning, the predominant approach, 195; natural engagement pattern, 220; removal of, 197; taking down, 216
Mission Impossible: Dead Reckoning, 150, 178
Mitchell, Margaret, 102
Mitchell, Melanie, 236, 238
models: base model, 133–34; language models, 134, 196–97; translation model, 191
Mollick, Ethan, 286
Mollick, Lilach, 286
Mona Lisa replication, 124–25
Mount St. Mary's University, 36–37
music industry, 21–22
Musk, Elon, 180
Myanmar, 187–89, 260
Myth of Artificial Intelligence, The (Suchman, Whittaker), 252

myths about AI: AI community, culture and history of hype, 235–39; AI hype, 231–35; AI hype, public figures spreading, 251–55; Bing chat, misleading news headlines, 249; cognitive biases, 255–57; Gartner hype cycle, 232; introduction, 227–31; news media misleads the public, 247–51; overview of, 32–33; reproducibility crisis, 241–47; robots featured in news media, 248; transparency, few incentives for, 239–41

Napalm Girl, 209–11
Narayanan, Arvind, 18–21, 62, 81, 234–35
National Center for Missing and Exploited Children (NCMEC), 194
National Eating Disorders Association (NEDA), 277
National Institute of Standards and Technology, 15
natural engagement pattern, 220, 221
Nature, 109
Netflix, 39
Netherlands, 48–49, 77, 85–86
neural networks: 5-layer neural network, 110; classify images of dogs, 119; failure and revival, 107–10; image generation, 121; ImageNet, 114; ImageNet competition, 112–13; text generation, 131. *See also* deep neural networks
NeurIPS, 237, 246
New York Magazine, 144
New York Times, 247, 248
news media, misleading the public: accuracy numbers, 248–51; AI hype, debunking, 247–48; civil war prediction, 250–51; clickbait, 248, 249; Epic, 248; headlines, 249, 250; hype, underlying reasons for, 251; images of AI, 247; robots in news media, 248; universities, press releases from, 250
next-word-prediction, 132–33, 248
Nimda, 176
No to AI Images, *126*, 126
noise, 121
nonproliferation, 174
notable historical computers, *157–59*
NVIDIA, 114

object recognition, 13
Ohio Risk Assessment System (ORAS), 51
Online Safety Bill (2022) (UK), 143
opaque AI incentivizes gaming, 46–48
opaque AI models, 46–48
OpenAI: AGI, 151; AI regulations, 274; ChatGPT, 3–5; Codex, 245; fine tuning, 135–36; public knowledge to trade secrets, shift from, 260; regulatory capture, 274; Sama, pay disparity, 144; toxic speech and offensive outputs, 269
optimization mindset, 266
Optum's Impact Pro, 54–55
Oregon experiment, 268
Ortiz, Karla, 290
Othello, 138
overautomation, 48–51
overreaching, 187, 197
Overton window of speech, 211, 212

paper clip maximizer, 2, 171
Papert, Seymour, 235
partial lotteries, 266–68, 288
the past, predicting, 194–97
peaks, (springs), 235

peer review, 242
peer-reviewed, 24, 228, 239, 242
perceptron, 105–7, *106*
Person of Interest, 75
Photoshop, 143
Pineau, Joelle, 246
Pitts, Walter, 105
policymaking, 212, 217, 218
political units, datasets, 91
power, defined, 171
PowerPoint presentations, 278
precarious work, definition of, 144
predicting the future: AI fails, five reasons why, 59; computers, predicting the future with, 62–67; COVID interventions, case study (New Zealand), 96; criminal justice, predictions in, 78–81; fragile families challenge, 67–70; fragile families challenge, disappointing end, 73–78; getting specific, 67–70; individuals to aggregates, from, 90–97; introduction, 60–62; limits to, 97–98; meme lottery, 86–90; predicting success, 81–86; prediction systems, 60; weather forecasting, 62–64, 65; wrong people, predictions about, 51–53
predictive AI: Allstate, 10–11; automated decision-making, 28–29; concluding thoughts, 58–59; failure, example of, 9–10; five reasons why AI fails, 59; gaming the system, 11–12, 47–48; good prediction is not a good decision, 43–45; in hiring, 10; in industry and government, 24; insurance rates, 10–11; introduction, 2, 3, 9–12, 36–38; life-altering decisions, making, 38–43; predicting the future, 29–30. *See also* the future, getting specific
predictive AI goes wrong: automated hiring tools, 18, 24; concluding thoughts, 58–59; EAB Navigate, 37; five reasons why AI fails, 59; healthcare decisions, 43–45; incentivizing gaming, 11–12, 46–48; inequalities, exacerbating existing, 53–55; introduction, 36–38; life-altering decisions, making, 38–43; overautomation, 48–51; randomness, 56–58; university retention rate, 36–37; world without prediction, 56–58; wrong people, predictions about, 51–53
presidential elections: 1948 presidential election, 56; 2016 presidential election, 199; *Chicago Tribune*, 1948 election, 56
pretraining, 135, 136, *163*
priming, 256
pro-ana posters, 221
products, overhyped, 24–25
programmatic interface, 175
prohibition, 275
proprietary, 228, 246–47
PSY, Gangnam Style, 88
psychohistory, fictional science of, 90, 91
Ptacek, Thomas, 3–4
public figures, spreading AI hype: Daniel Huttenlocher, 251; Eric Schmidt, 251; Future of Life Institute, 254–55; Henry Kissinger, 251
public interest, promoting AI: false dichotomy, cutting through, 268–74; introduction, 258–61; regulation, limitations of, 274–75; snake oil, appealing to broken institutions, 261–65

Public Safety Assessment (PSA), 51, 52–53
Pulitzer Center fellowships, 26
Pymetrics, 24–25

QR code, AI generated image, *100*
QR codes, 100–101
qualitative criteria, 68
quantification bias, 257

racial bias, 54, 79–80
radiologists, replacing, 238–39, 276
Rahimi, Ali, 237–38
randomized controlled trials (RCTs), 45
randomness: discomfort with, 56–58; election forecasting, 56–57; embracing, 265–68; housing allocation lotteries, 58; psychohistory, 90
Recht, Benjamin, 237
recidivism-prediction models, 239
recommendation algorithms, 31, 88, 211, 220–22
recommender systems, 260, 271
Reddit, 213, 224–25, 226
Reddit-style moderation, 224–25
regulation: automated notifications of copyright claims, *209*; content moderation decisions, 205–6; copyright, 206–9, *209*; fake copyright strikes, 208–9; hate speech, 272; horizontal regulations, 270, 271; limitations of, 274–75; overzealous regulation, 275; platforms have free reign, 205; social media, 205–9; U.S. copyright law, 208; vertical regulations, 270, 271
regulation: cutting through the false dichotomy: China, 271; companies, 268–69; enforcement, 273; environmental protection, 269; EU, 270–71; food safety, 269; labor rights, 269; lagging behind, 272; myths, 269, 272–73; regulation, definition of, 268; regulation, improvement of, 273; self-regulation, 271, 272; tech regulation, 272–73; United States, 270, 272
regulations: AI for surveillance, 129; annotators, 144; China, 271; Clean Air Act, 269; Clean Water Act, 269; Digital Services Act, 223–24; enforcement of, 273; false dichotomy, cutting through, 268–74; labor rights, 269; limitations of, 274–75; U.S. National Labor Relations Board, 279
regulatory capture, 274–75
Reid, Randall, 15
relative accuracy, 79, 228
reproducibility, definition of, 242
reproducibility challenge, 246
reproducibility crisis in AI research: AI-based science, 244–45; civil war prediction, 243–44; Codex, 245; commercial models, reliance on, 245; conclusion, 246–47; improving reproducibility, 245–46; leakage, 244; NeurIPS, 246; online workshop, 244; oversight, lack of, 231; overview of, 241–42; peer review, 242; reproducibility challenge, 246; scientific advances, genuine, 245; social psychology, 242; 2018 study, 243
resource scarcity, eliminating, 265, 268
Retorio, 47
reverse image search, 118–20, 127
rich-get-richer dynamics, 83–84, 86, 88, 266–67, 268
risk assessment tools, 40–41, 79–80
Rite Aid, falsely accusing customers, 17

River Dell High School students, 290
Robodebt scandal (Australia), 49
robot lawyer, 104
robot tax, 280
robotics, 32, 256
robots, 247, *248*, 256
rock-paper-scissors, 136–37
rogue AI: AI safety community, 170; AI safety meme, *170*; overview of, 168–71; paper clip maximizer, 168–69; power, definition of, 171; power-seeking behavior, 169; reframe the issue from intelligence to power, *171*; superintelligent beings, 171; us-versus-it argument, 168
Rohingya (Myanmar), persecution of, 188–89
Roose, Kevin, 101
Roosevelt, Franklin D., 275
Rosenblatt, Frank, 105, 153, 235
Rowling, J. K., *Harry Potter*, 82
rules-based system, 165–66
Russakovsky, Olga, 111
Russia, 199, 244

Sacco, Justine, 89, 222
Salganik, Matthew, 19, 62, 71–73, 76, 84–85
Sama, 144
San Diego's surveillance program, 265
Santa Clara Principles, 192
Schmidt, Eric, *The Age of AI*, 251, 252
schools and colleges, 8–9, 262, 265
Schwartz, Barry, 267
Science, 245
science self-corrects, 23
Scientific American, 21
scientific research, 242–47, 250, 267
screen time, 285
scriptwriting, 279
selection bias, 152–53
self-driving cars, 14, 32, 153, 168. *See also* autonomous driving
Sen, Ari, 26
sentient, 25, 101
sepsis, predicting the risk of, 227–29
Seuss, Dr., 82
shadow banning, 183
Shazam, 207
ShotSpotter, 263–64
Shutterstock, 123
simplicity, 266
simulation, 64–66
Simulmatics, 65
snake oil: AI hype vortex, 21–26; AI journalism, 251; Artificial General Intelligence, 152; broken institutions, appealing to, 261–65; cognitive biases, 257; commercial AI snake oil, 23; content moderation AI, 226; criti-hype, 178; definition of, 2, 26–34; investigating accuracy, 253; Mark Zuckerberg, 179; myths about AI, 230; predictive AI, 37, 57, 58, *59*, 98; robot lawyer product, 104
snake oil advertisement (1905), 27
social media: automated notifications of copyright claims, 209; content moderation, 194–97; context, taken out of, 183–88; coordinated mass harassment, 89; cultural incompetence, 188–94; drawing the line, 209–16; evading content moderation, 198–201; Facebook's moderator training materials, *196*; fingerprint matching, 194, 206, 218; introduction, 179–83; life and death, a matter of, 201–5; Overton window of speech, *212*; overview of, 31; the past, predicting, 194–97; problematic

social media (*continued*)
content, amplification of, 221; problems of their own making (platforms), 218–23; regulation, adding to the mix, 205–9; seven reasons content moderation is hard, 217; stages of AI-based content moderation cycle, 214. *See also* content moderation

social media companies, 32, 88, 147, 191, 216

social media, content moderation: content moderation, evading, 198–201; fingerprint matching, 194; the future of, 223–26; machine learning, 194–95; platform's failures, 218–23; seven shortcomings of AI, 216–18

social predictions, improving, 74, 75, 76

social science, 70–71. *See also* Fragile Families Challenge

Social Sentinel, 26, 264–65

South Korean government, 127

spam, 66, 181, 198

spam classifiers, 66

specific threats, defending against, 174–77

speech, regulation of, 32

springs (peaks), 235

Sri Lanka, 190, 260

Stability AI, 7, 134. *See also* Stable Diffusion

Stable Diffusion, 2, 7, 121–22, 122–23

Star Wars, 82–83

STAT, 230

Steinhardt, Jacob, Troubling Trends in Machine Learning Scholarship, 238

STEM majors, 37

stimulus checks, COVID-19, 38

stock photos, 104

Strange World, 83

students: AI-generated text, 262–63; ChatGPT, 9; cheating detection software, 33–34; exercises and pedagogical materials, 35; ImageNet, 111; introduction, 9; Limits to Prediction, 19–20; mental health support, 264; online course, Bitcoin and cryptocurrencies, 235; partial lotteries, 267; predictive AI, 36–37; Princeton, 8; River Dell High School, 290; Social Sentinel, 26, 264–65

success, predicting, 81–86

Suchman, Lucy, The Myth of Artificial Intelligence, 252

suckers list, 10

suicide, a matter of life and death, 201–5

suicide prevention efforts, 201–3

superintelligence, 151, 166, 171

superintelligent beings, 171

Support Vector Machine (SVM), 109–10

Suri, Siddharth, *Ghost Work*, 278

surveillance, 127–29

Sussman, Gerald, 153

Sutskever, Ilya, 112–13

SVM, 109–10, 112, 114

symbolic systems, 108–9, 160, 161

system dynamics, 64

systems thinking, 224

tech journalism, 26, 284

teen mental health, 285

Telangana (India), 127

Telegram app, 142–43

Test of Time Award, 237–38

Tetlock, Philip, 154

text generation, 129–33

text-to-image generation: already-deployed models, harms of, 254;

appropriating creative labor, 122–23; Dall-E 2, 7; diffusion model, 121–22; entertainment, 7–8; Firefly, 7; Midjourney, 7; Mona Lisa, 124; Stable Diffusion, 7

text-to-video generators, 8

threats (specific), defending against, 174–77

threshold of usefulness, 166

Tigray, Ethiopia, 189, 191

TikTok, 86, 88, 218–19

tobacco companies, 239, 274

token, 132, 133

toxic content, 143–44

training data: appropriating creative labor, 122–23; asthmatic patients, 43–44; chatbots, 5; content-moderation classifiers, 184–85; cost of improvement, 144; creating, 35; from images to text, 129, 132; leakage, 244; machine learning models, 66; from models to chatbots, 136; Ohio Risk Assessment System, 51; Optum's Impact Pro, 54; Public Safety Assessment, 52

training data, generative AI: biases and stereotypes, 103; competitions, 112; Google, 114; ImageNet, 115; Telangana, 127

translation tool, 134

transparency: Artificial Intelligence Act, 271; China, 271; content moderation, future of, 223–24; Digital Services Act, 270; language models, 141; OpenAI, 274; ShotSpotter, 264

transparency, incentives for: accuracy measurements, 240–41; benchmark datasets, 241; COMPAS, 239; early-stage startups, 239–40; Epic, 239; GPT-4 (OpenAI), 241; Hire-Vue, 239; top-N accuracy, 240; venture capitalists, 240

Troubling Trends in Machine Learning Scholarship (Lipton, Steinhardt), 238

truth of statements, evaluating, 196, 197

Turchin, Peter, 91

Turing, Alan, 156

Turing Award, 109, 164

Turkey, 216

Twitter: account suspension, unaccountable, 213–14; bridging-based rankings, 220; content moderation, 180; content moderation mistake, 184; meme lottery, 87; Turkey, blocking opposing voices, 216

20th Century Fox, 82–83

2023 Hollywood strikes, 7–8

Uber, 213

United Artists, 82

United States: Epic sepsis model, 227; false arrests due to facial recognition, 15; Social Sentinel, use in schools, 26

United States Postal Service, 109

United States, predictive AI: Affordable Care Act, 53; criminal risk prediction systems, 51–52; election forecasting, 56; racial disparities in policing, 55

United States, promoting public interest: AI for detecting weapons, 263; AI regulation, 270; community networks, 259–60; environmental protection, 269; gun violence, 263; internet connectivity, control of, 259; prohibition (alcoholic beverages), 275; regulation enforcement, 273; taxing AI, 280; unemployment insurance, 280

United States, social media: content moderation, 205–9; cultural incompetence, 192; DARPA, 259; luck, role of, 82; misinformation, 193; suicide/self-harm, 201–5
UnitedHealth, 50
Universal Basic Income (UBI), 279–80
Universal Pictures, 82
University of Chicago, 250
unpredictable events, 75–76
Upstart's model, 42
U.S. copyright law, 123, 208
U.S. Federal Trade Commission (FTC), 25, 273
U.S. National Labor Relations Board (NLRB), 279
U.S.-centric policies, 192
us-versus-it argument, 168
Utah (Live Time), 230
Uyghur population, 127

valleys, (winters), 235
vector, 119–20, *121*
vector similarity, illustration of, *121*
venture capitalists, 234, 240
Verizon, 259
village idiot, use of term, 170
Vinsel, Lee, 253
viral hits, 86–90
virality, 86–90
von Neumann, John, 63

Wang, Angelina, 24, 42–43
Wang, Rona, 103
WatchMojo, 208, *209*
watermarks, 123
Watkins, Elizabeth, 239
Watson, Emma, 142
weather forecasting, 62–64, 65
weather prediction, 67–68

web scraping, 115, 191, 289
Web3, 233–34
weights, 106–7
Weizenbaum, Joseph, 165
welfare checks, 203–5
WhatsApp, India, communal violence, 190
White, E. B., 82
White, Molly, 234
White House executive order on AI, 270
Whittaker, Meredith, *The Myth of Artificial Intelligence*, 252
Williams, Adrienne, 146
Williams, Robert, 15
Winecoff, Amy, 239
winters (valleys), 235
Woodruff, Porcha, 15
work, AI and the future of, 276–81
World Wide Web, 233
worms (viruses), 176

X (formerly Twitter): account suspension, unaccountable, 213–14; bridging-based rankings, 220; content moderation, 180; content moderation mistake, 184; meme lottery, 87; Turkey, blocking opposing voices, 216
x.ai (calendar scheduling company), 230

YouTube: Charlie Bit My Finger, 86–87; Content ID, 206–9; content moderation mistakes, 184; copyright, exception to 1996 law, 206; copyright strikes, fake, 208–9
Your Face Belongs to Us (Hill), 16

Zeihan, Peter, 91
zero-day vulnerabilities, 175
Zuckerberg, Mark, 179, 220, 221